Contents

A Galaxy Far Away

Matthew Usher

KALMBACH BOOKS

Dedication

For my parents, Wade and Mary Ann, and my wife, Cathy

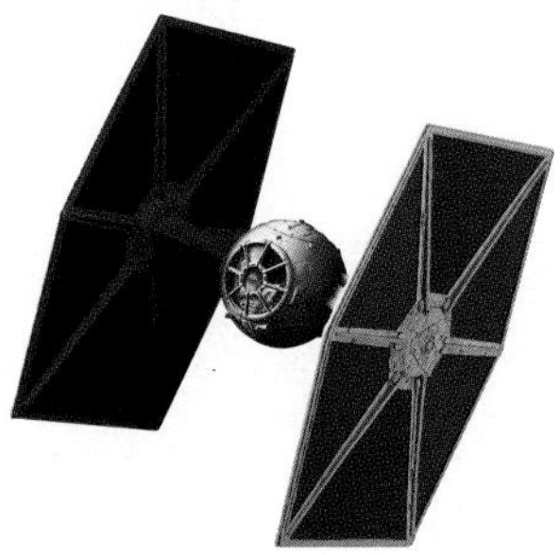

 Published by Kalmbach Publishing Co., 21027 Crossroads Circle, Waukesha, WI 53187.

Printed in the United States of America

02 03 04 05 06 07 08 09 10 11   10 9 8 7 6 5 4 3 2 1

Visit our website at
http://kalmbachbooks.com
Secure online ordering available

Publisher's Cataloging-in-Publication
(Provided by Quality Books, Inc.)

Usher, Matthew.
A galaxy far away : building Star wars models /
Matthew Usher. — 1st ed.
p. cm.
ISBN 0-89024-405-7

1. Space ships—Models—Design and construction.
2. Space vehicles—Models—Design and construction.
3. Star wars films. I. Title.

TL844.U84 2002   629.47'022'8
QBI02-200212

Art director: Kristi Ludwig
Book design: Anthony Belli

Foreword

Aircraft, tank, and car modelers almost always have a steady supply of new, state-of-the-art kits to build. If what's in the kit box isn't enough, they can turn to any one of several aftermarket companies that will supply all the add-on detail parts they could ever need. If you want to build a P-51 Mustang, detailed kits are available in several scales. There is also a huge variety of decals, multimedia detail sets, figures, and research materials that will help you build it exactly the way you want it with little trouble.

That's unfortunately not the case for science-fiction modelers. In most hobby shops, the section of science-fiction kits is usually quite a bit smaller than the sections of aircraft and tank kits. And even with the large number of *Star Wars* kits that have been produced over the years, there's been only one A-Wing kit and one Imperial AT-AT kit. If you want to add extra detail to either of them, you'd better figure out a way to do it on your own. With few exceptions, there are no aftermarket detail parts to help you.

That's not always a problem, though. Science-fiction modelers (*Star Wars* fans among them) are a resourceful, clever bunch, and for decades we've been finding ways to build, detail, and even light our models in an effort to make them look just like their on-screen counterparts. I've seen astounding, superdetailed models built from simple snap-together kits, and in almost every case that extra detail started life in the mind of a dedicated, imaginative modeler.

I've tried to use my own enthusiasm and imagination on each of the projects in this book, and hopefully it'll pass on to you. I tried to pick a good cross-section of kits from the *Star Wars* catalog, and I've tried to find a way to add just a little bit more to each model, beyond what came in the kit box. None of the projects is terribly exotic. I tried to use off-the-shelf paints and supplies wherever possible. Most of the electronics and lights I've added have straightforward circuits powering them. I hope each project will offer a little inspiration and start you thinking about how you can improve your next one.

Let's get started!

Trade Federation Droid Fighter

ONE

Built straight out of the box, AMT/Ertl's snap-together Trade Federation Droid Fighters kit (no. 30118) will take you less than 10 minutes to assemble. The box includes three, four-piece fighters and a four-piece display base. It's a disappointing combination at first glance.

Looking a little closer reveals the kit's strengths. First, although they're extremely simple and less than 3" long, the Droid Fighters are nicely molded, well detailed, and look accurate compared to photos of the studio models. Second, they are the same scale as the Naboo Fighter kit, which started me thinking about display options.

Examining the kit, I decided to build just one of the Droid Fighter models, and detail it as much as possible with an accurate paint scheme and a little weathering. I'd make a new display base, too.

The most striking feature of the Droid Fighter is its glowing red sensor eyes. I decided to find a way to illuminate the eyes from inside the model. Like the TIE Fighter model elsewhere in this book, it would require installing the lighting components in the tight confines of the model and designing a new, smaller display base in which to hide the batteries and power switch. I headed to the workbench and started plotting.

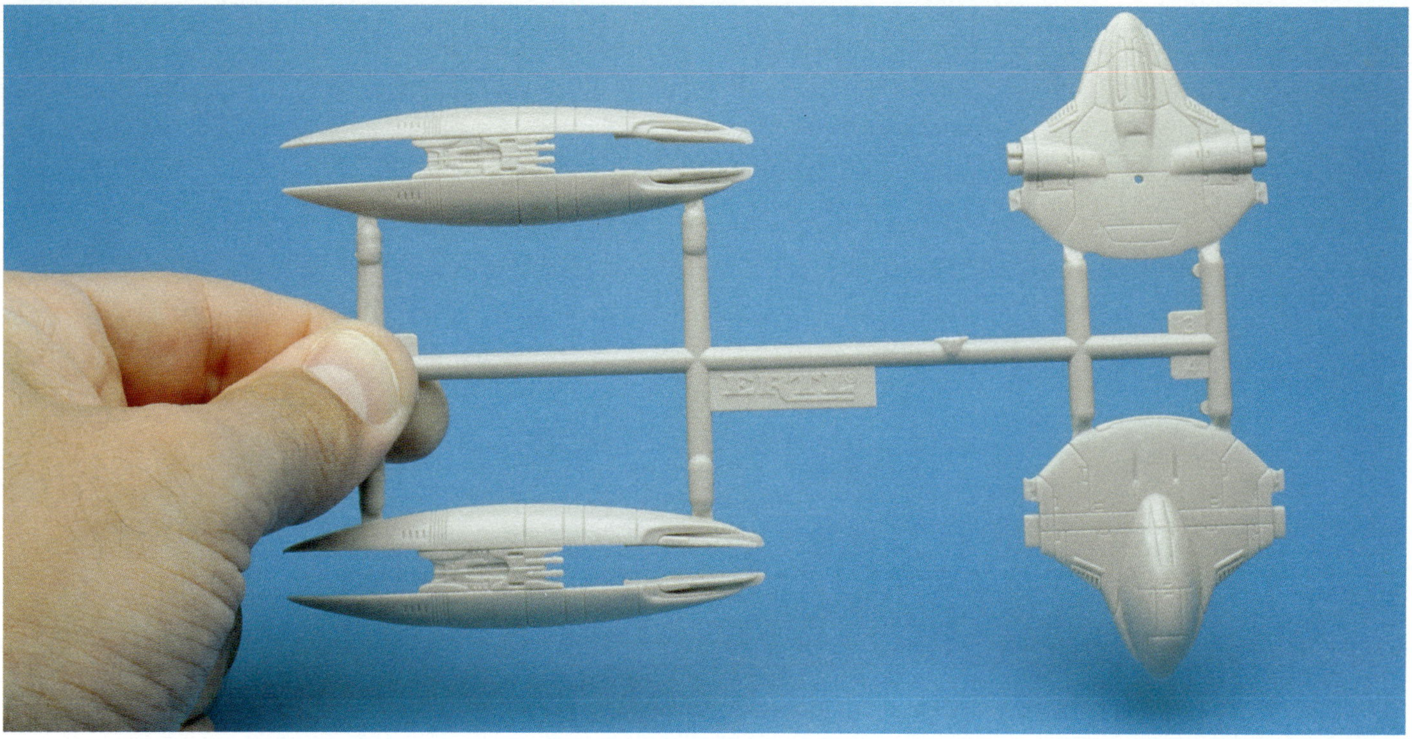

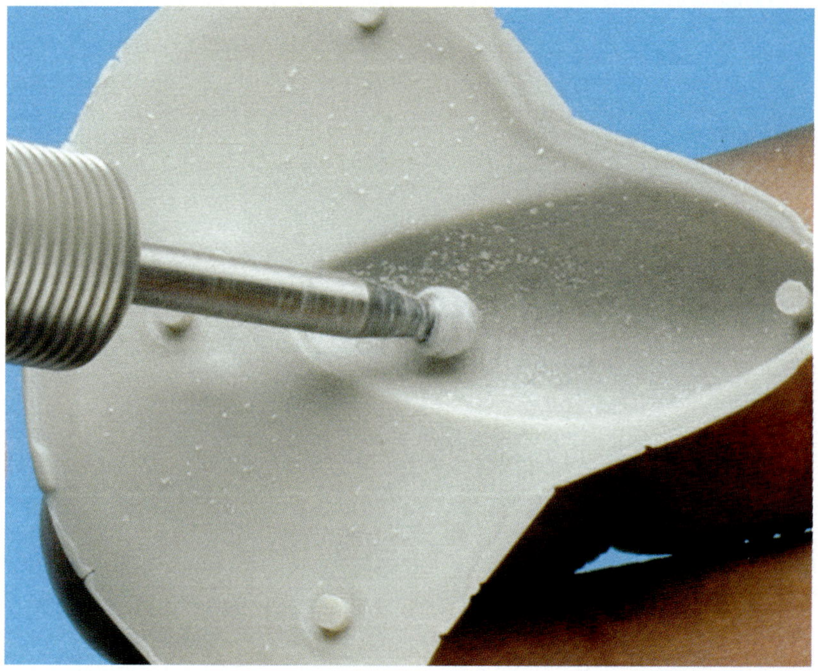

1 Clearly aimed at the beginner market, the snap-together Droid Fighter kit contains three four-piece Droid models and a simple display base. This model will need a new display base to support the completed in-flight model and to hide the batteries required for the light that will illuminate the "eyes."

2 Before cutting the eyes open, I thinned the area behind them with a steel cutter chucked in a motor tool. Use the motor tool's slowest speed, and check your work often. You want to remove as much plastic as possible without cutting through. Always wear eye protection when using power tools.

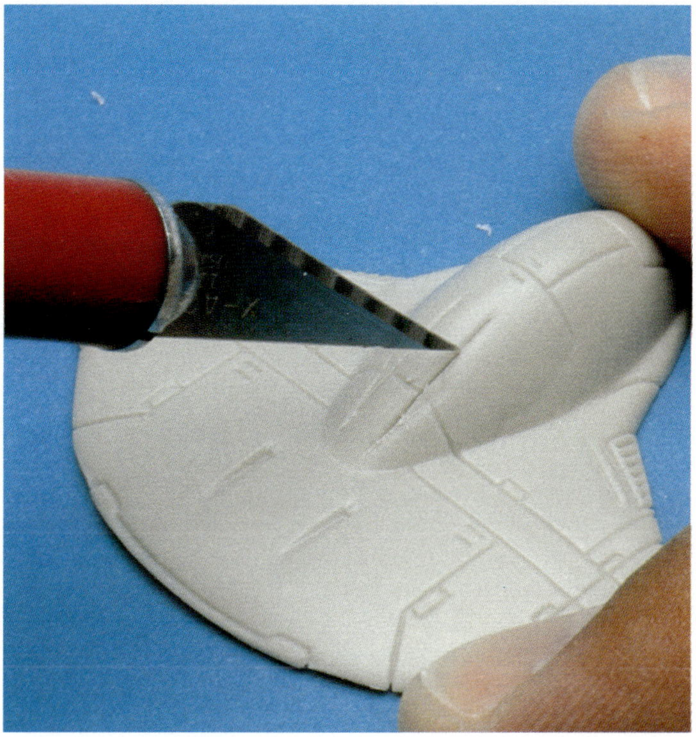

3 I used the back edge of a hobby knife blade to cut open the eyes, using their recessed edges as a guide for the blade. Each pass will remove a little more plastic until you cut through.

4 A thin strip of medium-grit sandpaper stretched between my fingers sharpened up the edges of the openings.

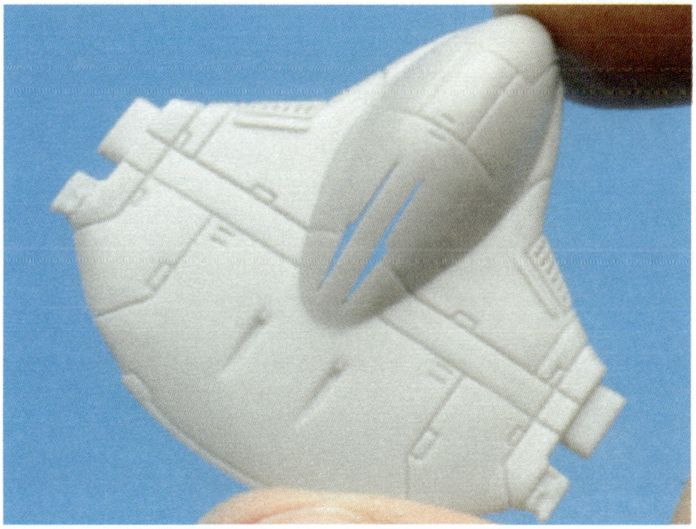

5 The finished openings have realistically thin edges.

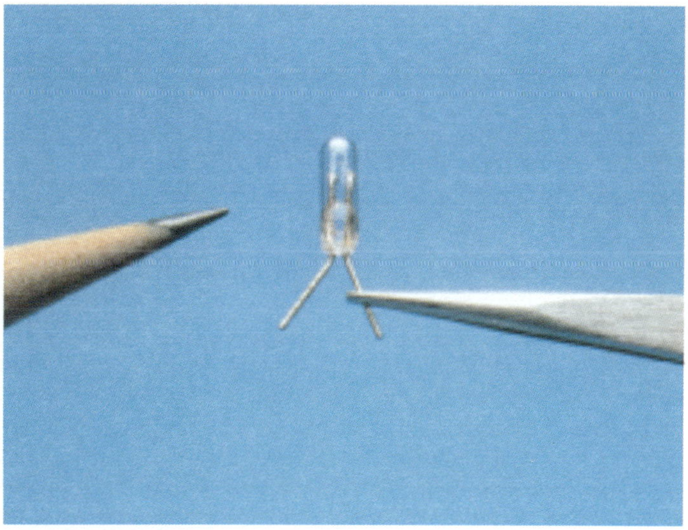

6 Instead of using a red light-emitting diode to light the eyes, I decided to use an incandescent krypton bulb from a pocket flashlight. A red gel glued above the light inside the model would give the light the proper color.

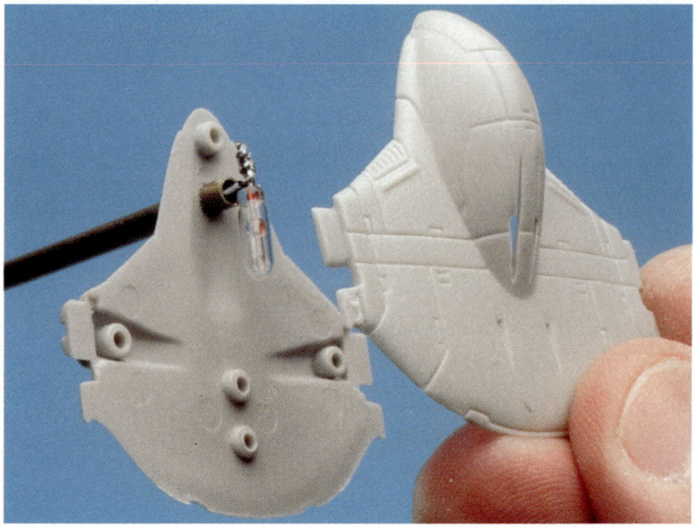

7 Powered by two batteries hidden in the base, the krypton bulb would be bright enough to light the eyes but small enough to fit inside the model. I soldered the leads to a piece of two-conductor model railroad wire. A piece of $\frac{1}{8}$"-diameter brass tubing would support the model and hide the wiring.

8 To put the light in the proper location inside the model, I drilled a $\frac{1}{8}$" hole in the rear of the baseplate for the support, which is made of brass tubing.

9 A test of the wiring revealed a big problem—the ultra-bright bulb lit up the entire model, not just the eyes!

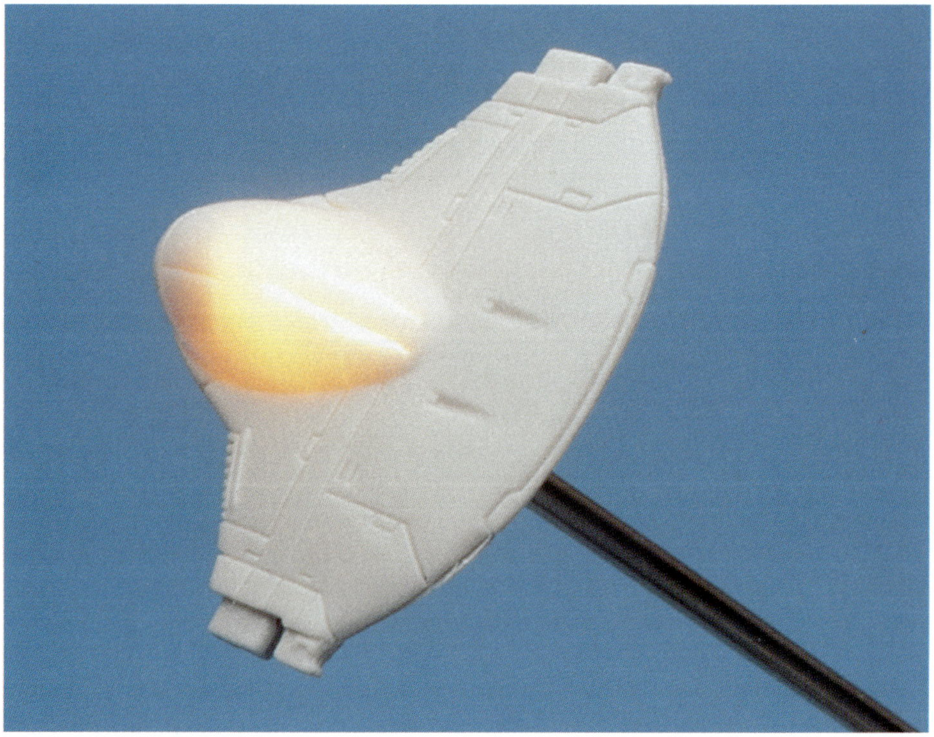

10 Normally, I'd paint the inside of the model silver to keep the light from shining through, but the krypton bulb was so bright I didn't think a simple coat of paint would be enough. I found the solution in my model car supplies. Bare-Metal foil is a thin, self-adhesive metal foil normally used to simulate chrome trim on car models. It would make the model's shell opaque. I applied the foil to the inside of the model in small patches, overlapping them slightly. Afterward, I cut the foil away from the eye slits with a sharp hobby knife blade, cutting from inside the model.

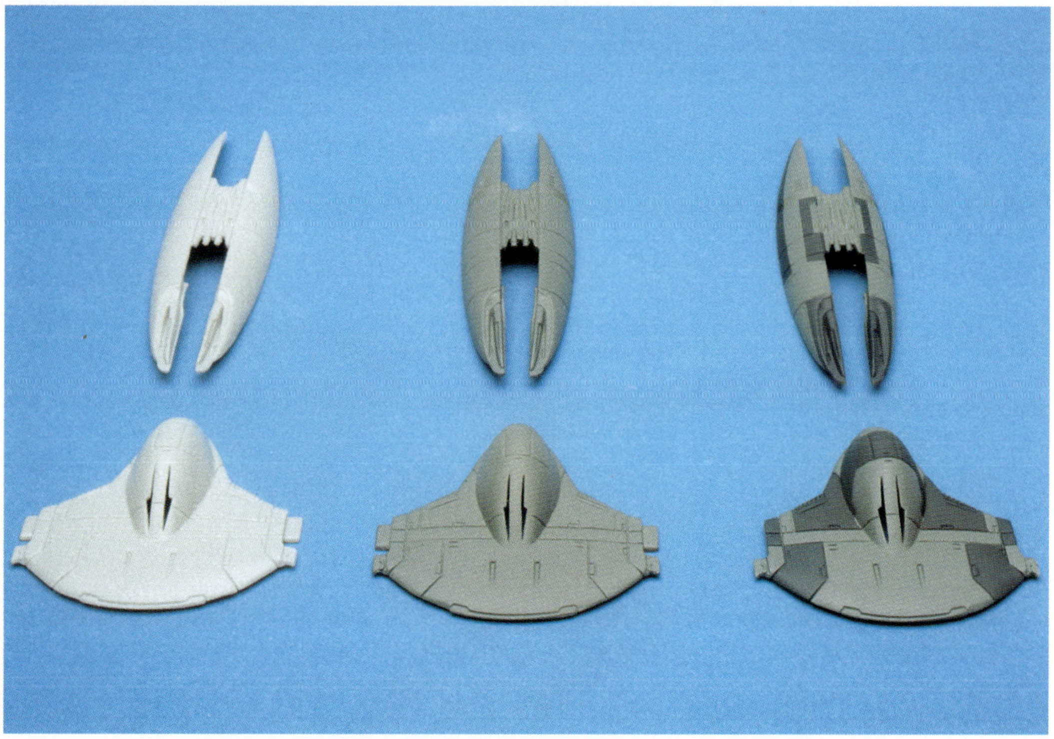

11 The model was ready for paint; I applied Testor Model Master enamels using an airbrush. First, I applied an overall coat of Faded Olive Drab (no. 2051). When it was dry, I masked the model and painted the darker sections with Green Drab (no. 1787).

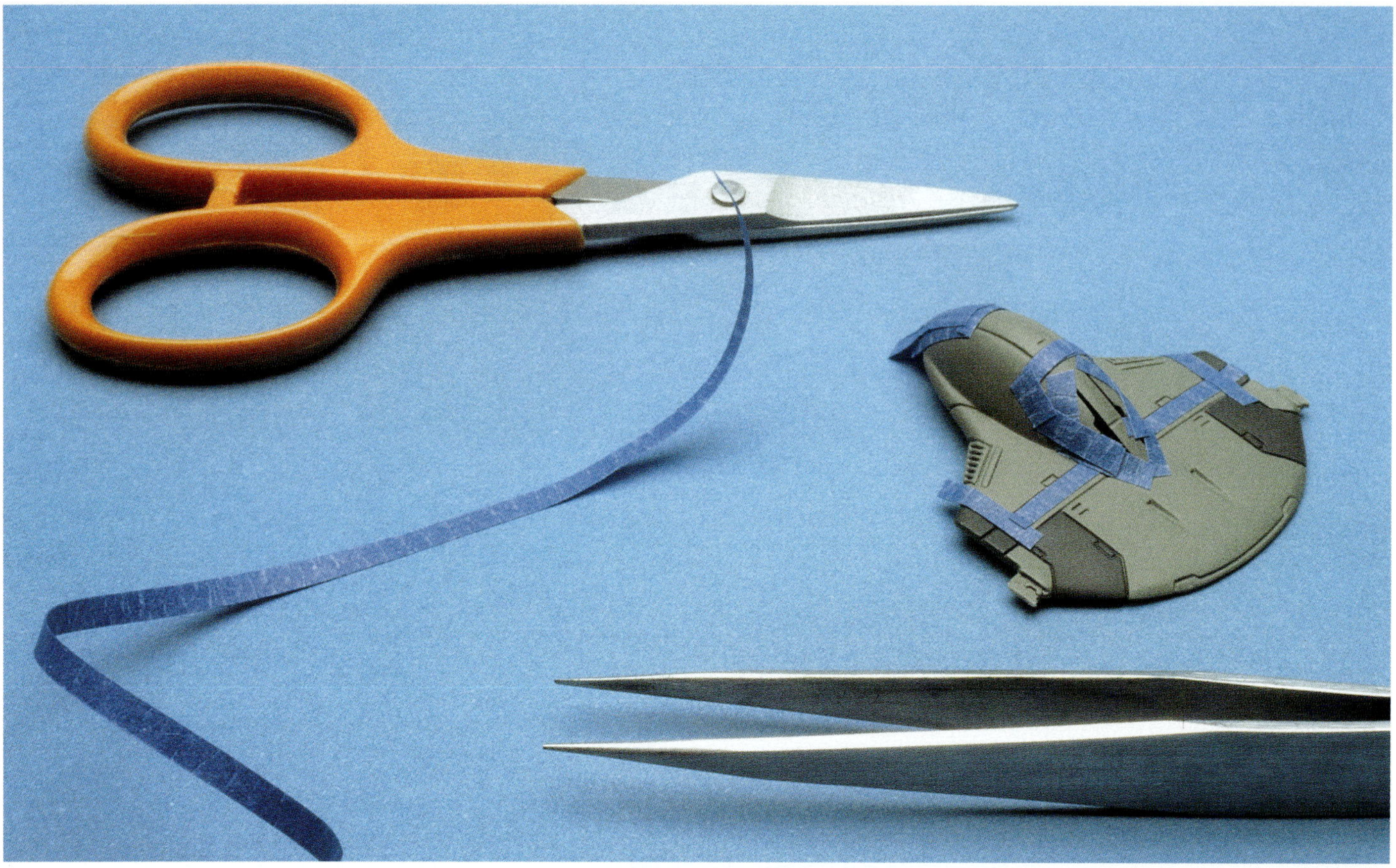

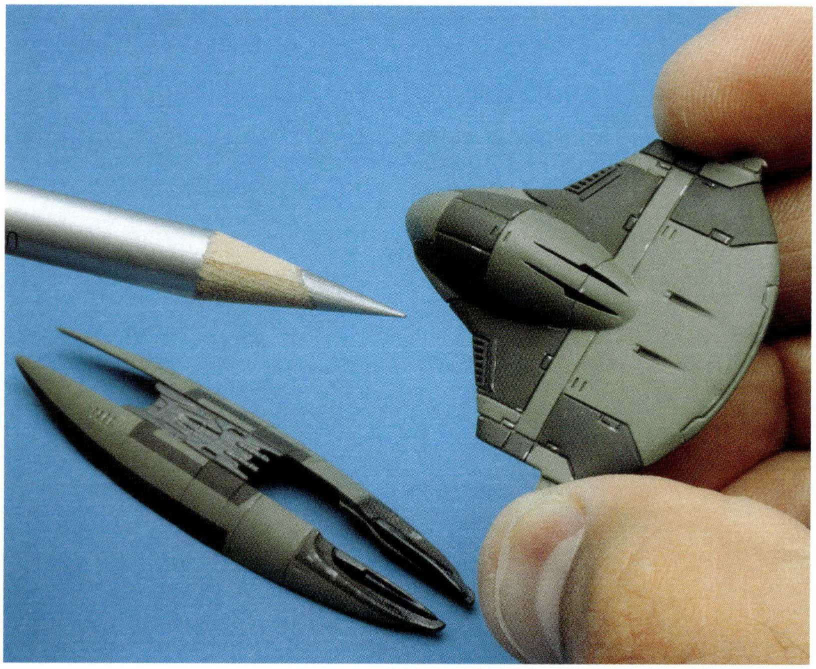

12 When the first coat was dry, I cut blue masking tape into thin strips to mask the model. I used the illustrations in *Star Wars: Episode 1 Incredible Cross Sections* and the box-art photo as guides for masking and painting the two-color scheme. I painted the darker sections Green Drab (no. 1787).

13 The Droid Fighters show a lot of wear and tear. I used a silver pencil from an art supply store to simulate chipped paint along the model's panel lines. I ground up a little dark gray pastel and applied it with a fine brush to simulate blast streaks.

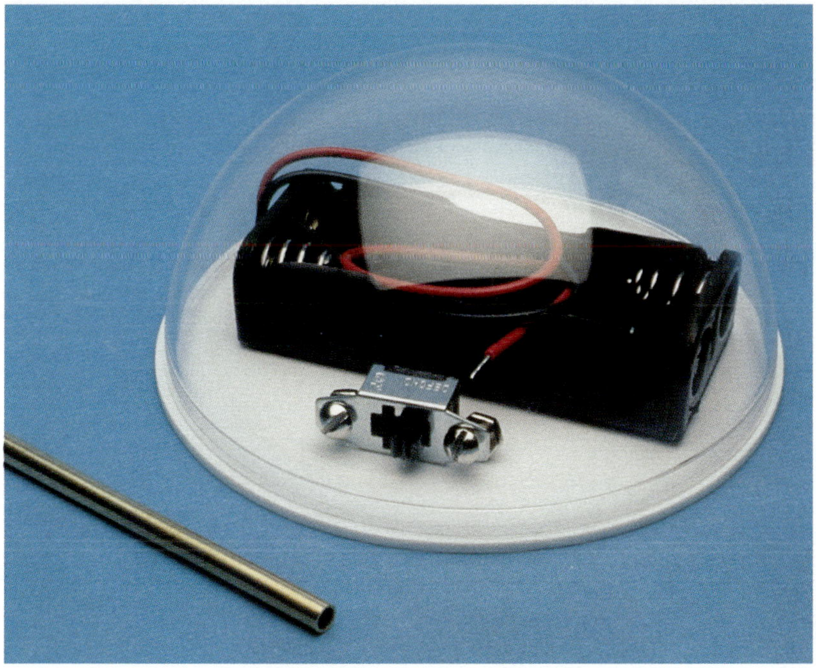

14 I glued the model's baseplate to the brass tubing with super glue. When it was dry, I added the rest of the parts, snapping the model together without glue. If I need to replace the bulb later, I can snap the model apart without damaging it. The red gel is just visible behind the model's eyes.

15 I found a clear plastic snap-together sphere in a craft store. One of the halves would work nicely as the base. Test-fitting showed that a AAA battery holder and slide switch would fit inside. I cut a round bottom plate for the base from .030″ sheet styrene.

16 The battery holder and switch are held in place with tiny nuts and bolts; pieces of styrene strip keep them aligned properly. I used a drill press to cut a ⅛" hole in the top of the plastic dome.

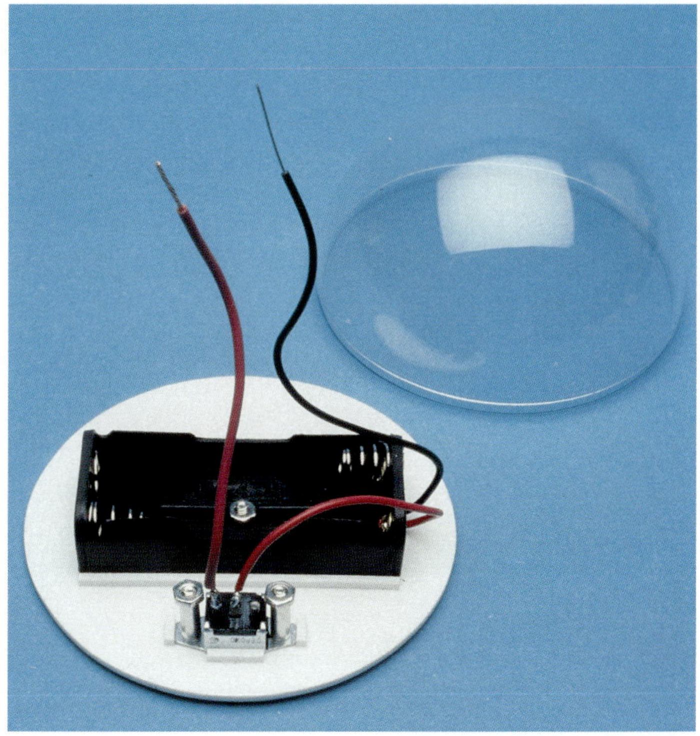

17 Before installing the electrical components permanently, I gave the base and dome a coat of Tamiya Gloss Black spray paint (no. TS-14). Make sure the paint is thoroughly dry before you proceed.

18 Finally, add the model to the base and hook up the wiring. You can attach the dome to the baseplate with tiny dots of super glue, but I wanted to find an easier way to access the batteries later. I glued two magnets to the base, and glued two steel washers inside the dome. The magnets stick to the washers and hold everything together securely.

TWO

Imperial Shuttle Tydirium

Released by MPC/Ertl with the premier of *Return of the Jedi,* the Shuttle Tydirium (kit no. 8733) is one of the better *Star Wars* kits of the era. While many of the kits released with the film are smaller, snap-together models, the Tydirium is a nicely detailed glue-together kit. It's a large model, too, rivaling only the Millennium Falcon for the amount of shelf space it requires. With its hinged wings deployed in flight position, the Tydirium has a 20″ wingspan and must be mounted on a display stand. To make the model easier to display, I decided to build the ship as it appears after landing, with its wings folded in their more compact vertical position.

In the film, Imperial Shuttles are shown with white formation lights in the vertical fin, wings, and lower fuselage. I decided to add these lights to give the model a little more visual interest and help break up its monochromatic paint scheme. The model has very little interior space, so instead of trying to hide batteries inside the fuselage, I decided to power the model externally with a model railroad transformer. A single two-conductor power line enters the model through one of the landing gear bays and branches out to each of the lights.

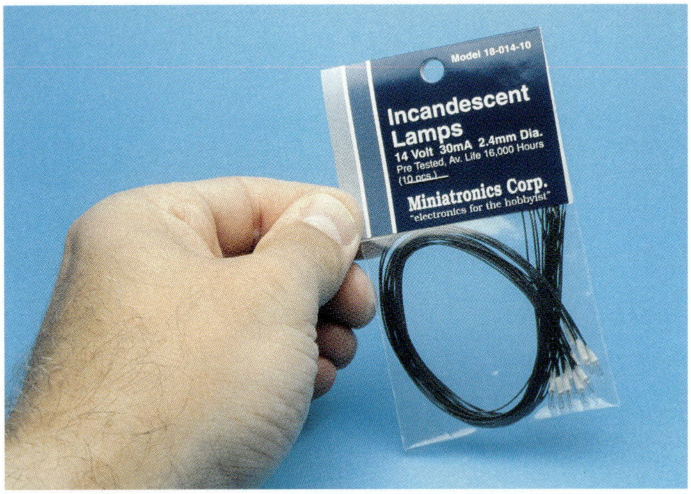

1 The Tydirium's formation lights should be bright white, so incandescent bulbs are the best choice. A trip to the model railroad section of the hobby shop turned up Miniatronics 14-volt 2.4mm bulbs (no. 18-014-10).

2 Miniatronics bulbs are the right size for the ship and have an estimated life of 16,000 hours.

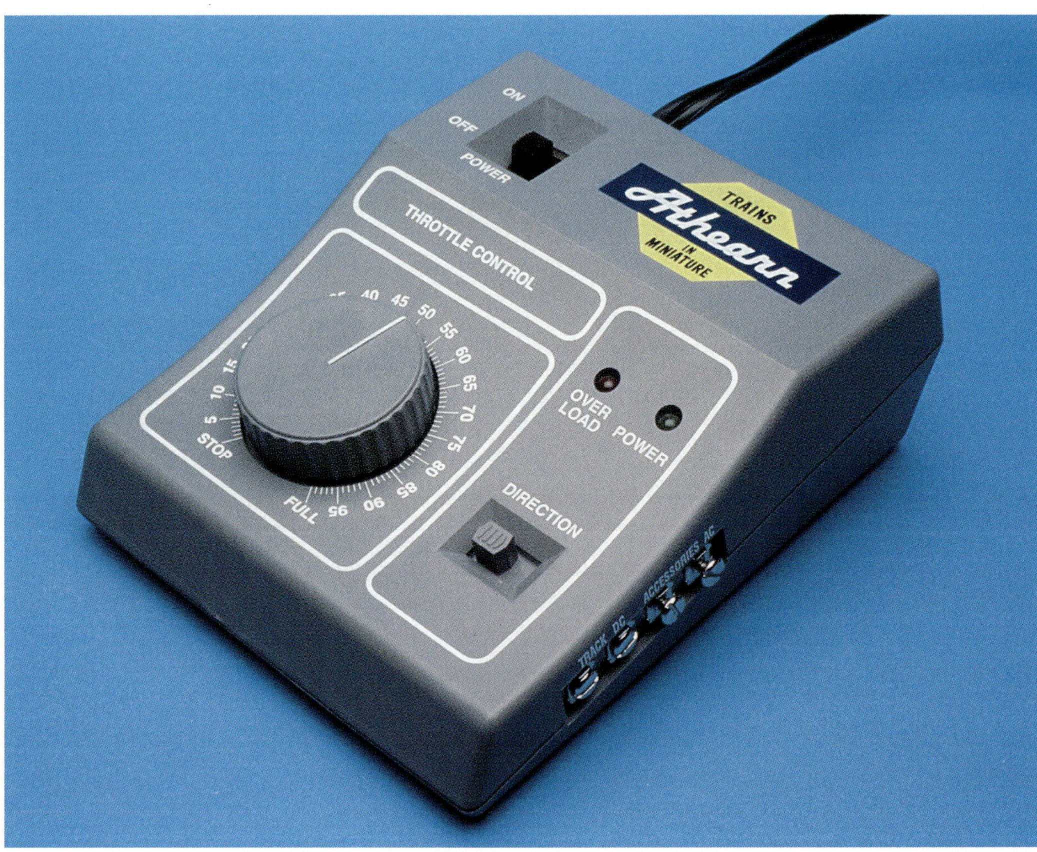

3 Instead of hiding batteries inside the model, I decided to use a model railroad transformer to power the model's lights. Simple transformers like this one come packaged with entry-level train sets and often end up in a drawer when they're replaced by more advanced transformers. If you have friends who enjoy model railroading, ask if they have a transformer like this one that they can part with. The speed control will enable you to adjust the amount of power going to the model.

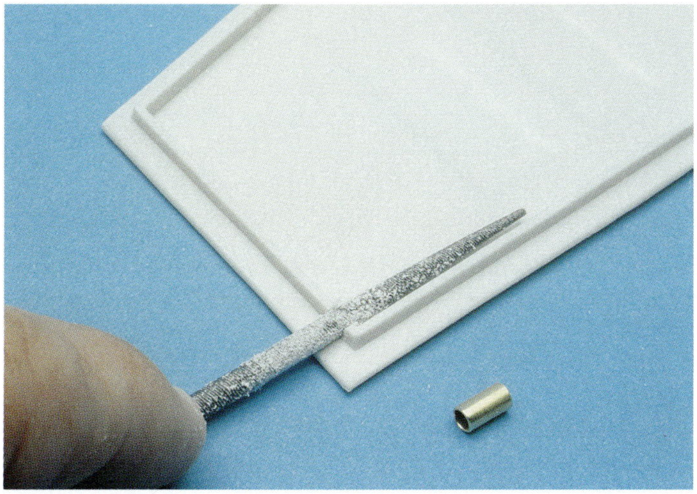

4 I decided to mount the bulbs in ⅛"-diameter brass tubing ferrules. They would make the lights easier to mount and insulate the surrounding plastic parts from the bulbs' heat. I used a round file to cut mounting holes for the ferrules in the wings and vertical fin. The panels are hollow and have plenty of room inside for wiring.

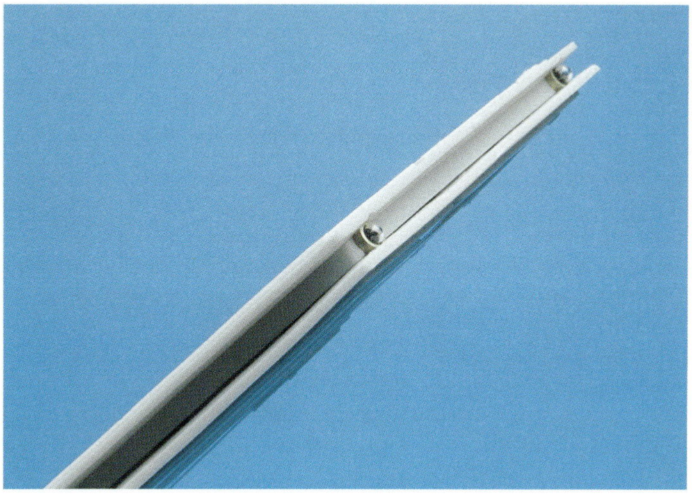

5 After a little filing and test-fitting, the ferrules fit nicely between the outer panels on the wings and vertical fin.

6 To keep the lights from shining through the vertical fin, I applied a strip of Bare-Metal foil to the model on both sides of each light.

7 I applied strips of Bare-Metal foil around the wing lights, too. The foil is opaque and extremely thin; it takes paint well.

8 The Tydirium should have a variety of mechanical details sandwiched between the outer panels of the wings and vertical fin; these details are easy to see on the boxtop photo. None of this detail is included on the model, but it's easy to add. I used several sizes and lengths of Evergreen Scale Models styrene strip and tubing to model the mechanical details, installing it with super glue.

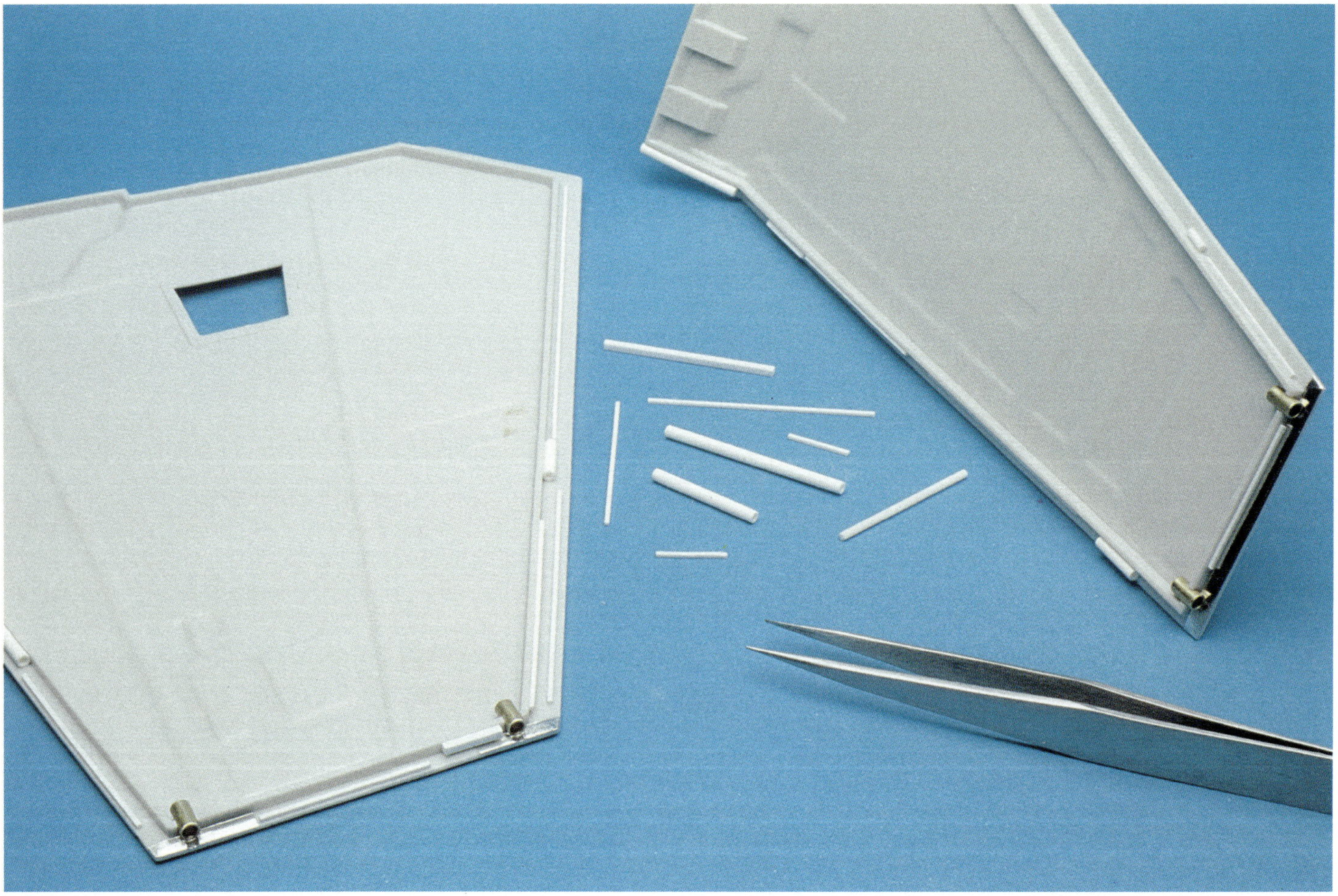

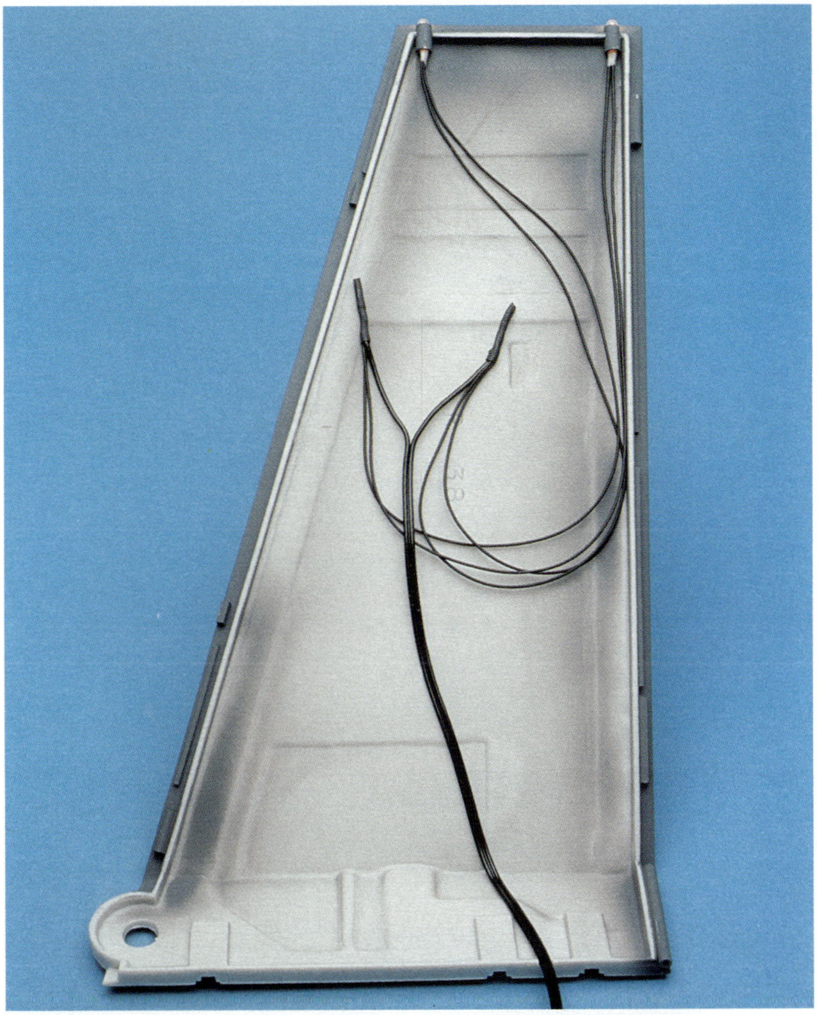

9 Here's a look inside one of the wings. Before closing it up, I soldered the lights' leads to a length of two-conductor model railroad wire and insulated the joints with Miniatronics heat-shrink tubing (no. HST-10). The wire runs out through a slot I filed in the inboard end of the wing. The lights are held in place inside the ferrules with a tiny drop of white glue. Should one of them burn out, I can pull it out with tweezers and splice in a new one without having to crack open the model. I airbrushed the edges of the panels dark gray at this point.

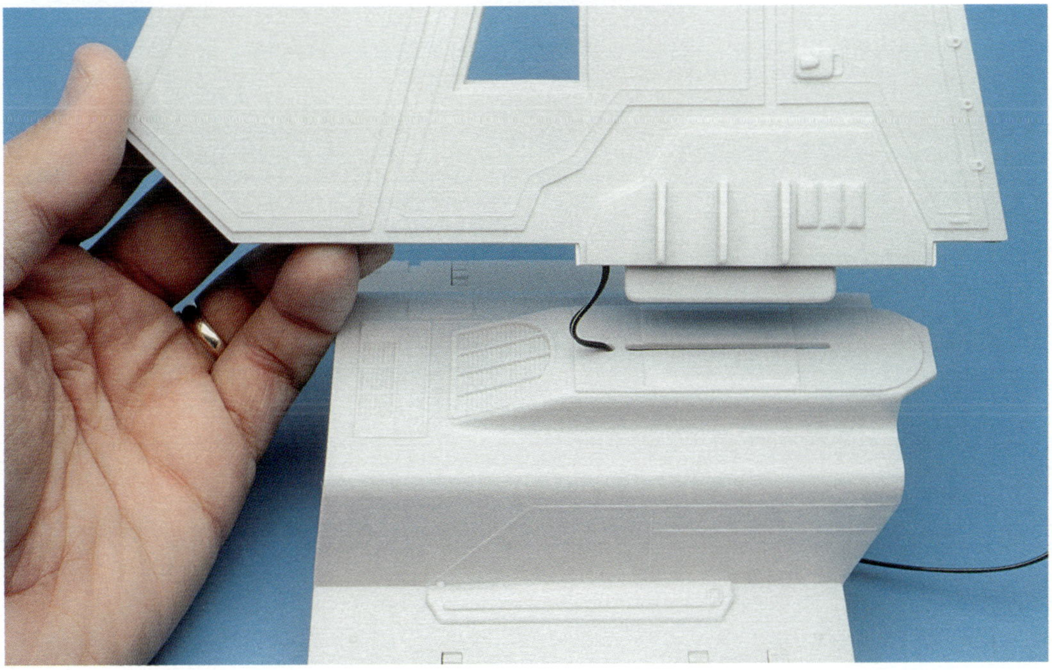

10 The wire from the vertical fin's lights runs out the bottom of the panel and into the main fuselage through a hole I drilled just in front of the mounting slot. The wires from the wing lights enter the fuselage through similar holes.

11 Each side of the vertical fin has a separate panel that must be glued in place. I filled the seam around each panel with gap-filling super glue. When it was dry, I wet-sanded the area until it was smooth. I made sanding blocks by gluing 400-grit wet-or-dry sandpaper to rectangles of 1⁄8"-thick balsa sheet.

12 Later versions of the Tydirium kit include crystal-clear canopies. While this is usually a good thing, photos of the studio model show a heavily tinted, almost opaque canopy.

13 To simulate the look of the studio model, I airbrushed the inside of the canopy flat black.

14 Moving to the main fuselage, I drilled holes for the two ventral lights and installed a pair of brass-tubing ferrules.

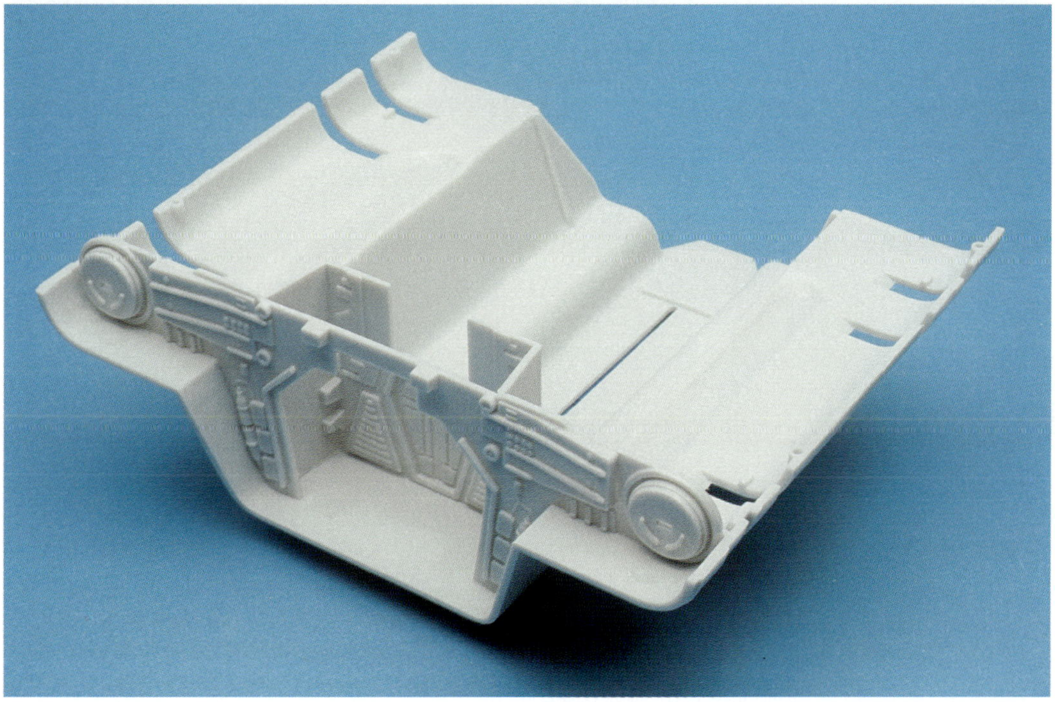

15 The upper half of the fuselage in my kit was slightly warped; installing the forward bulkhead parts helped correct the problem. I added fillets of gap-filling super glue along the joints inside the model where they wouldn't show. The fuselage is the backbone of the model, so it should be as strong as possible.

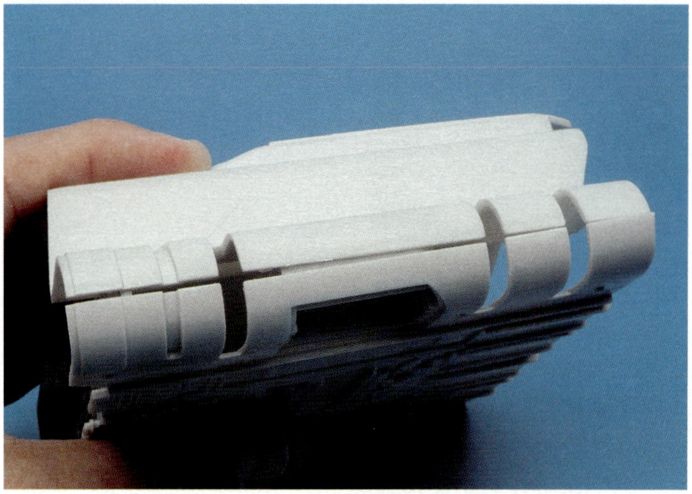

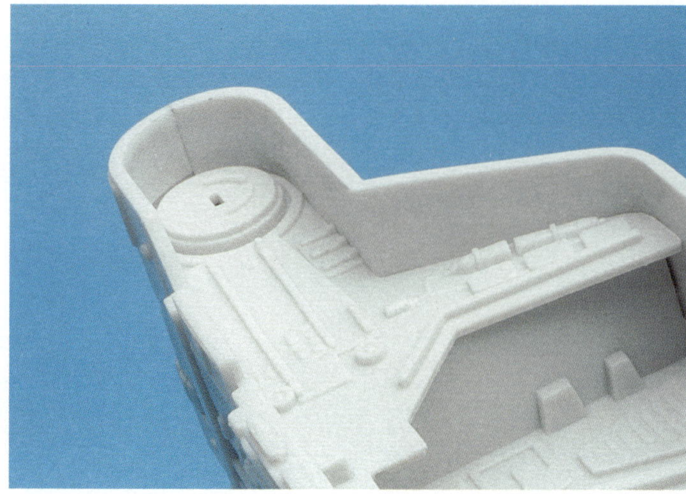

16 With the wings in their folded position, the seam between the upper and lower fuselage halves will be clearly visible. Carefully clean up the wing-hinge slots with a sanding stick before installing the hinges and gluing the fuselage halves together.

17 The seam on the inside surface of the hull is visible and would be tough to fill with putty or gap-filling super glue.

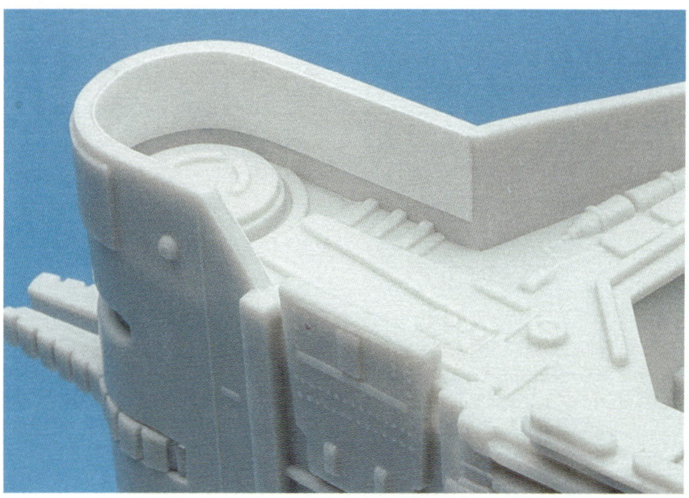

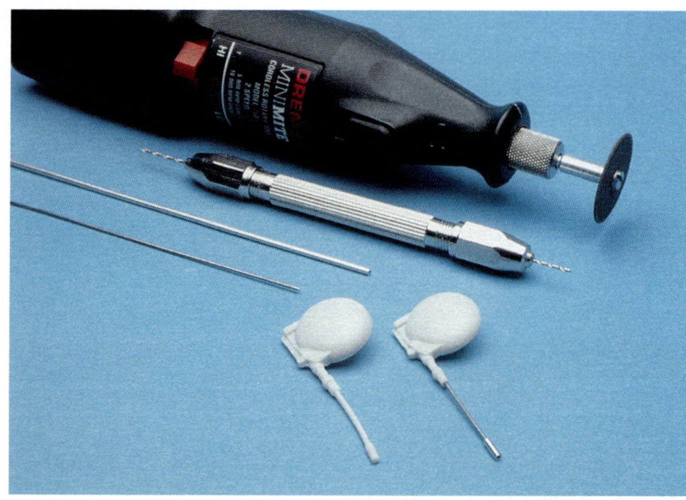

18 Instead of filling the seam, I covered it with a strip of .010"-thick sheet styrene. Gap-filling super glue blended the edge of the sheet into the edge of the hull.

19 The Tydirium is bristling with laser cannons, and they're a real focal point of the model. Unfortunately, the cannons included in the kit are poorly detailed, and most of the gun barrels are warped. I cut off the kit's gun barrels and replaced them with new ones I made using music wire and aluminum tubing. Here's one of the kit's wing-mounted cannons on the left, and the improved version on the right. Wear eye protection if you use a motor tool to cut the music wire.

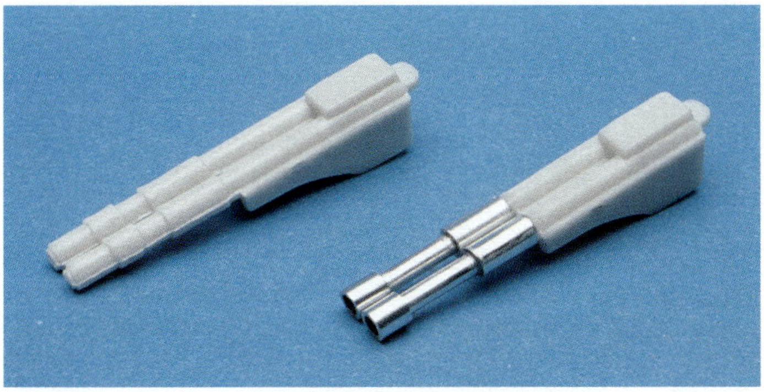

20 The forward-firing fuselage-mounted guns need some work, too. Again, I cut off the plastic barrels and replaced them with new ones I made from aluminum tubing. Gap-filling super glue holds everything together.

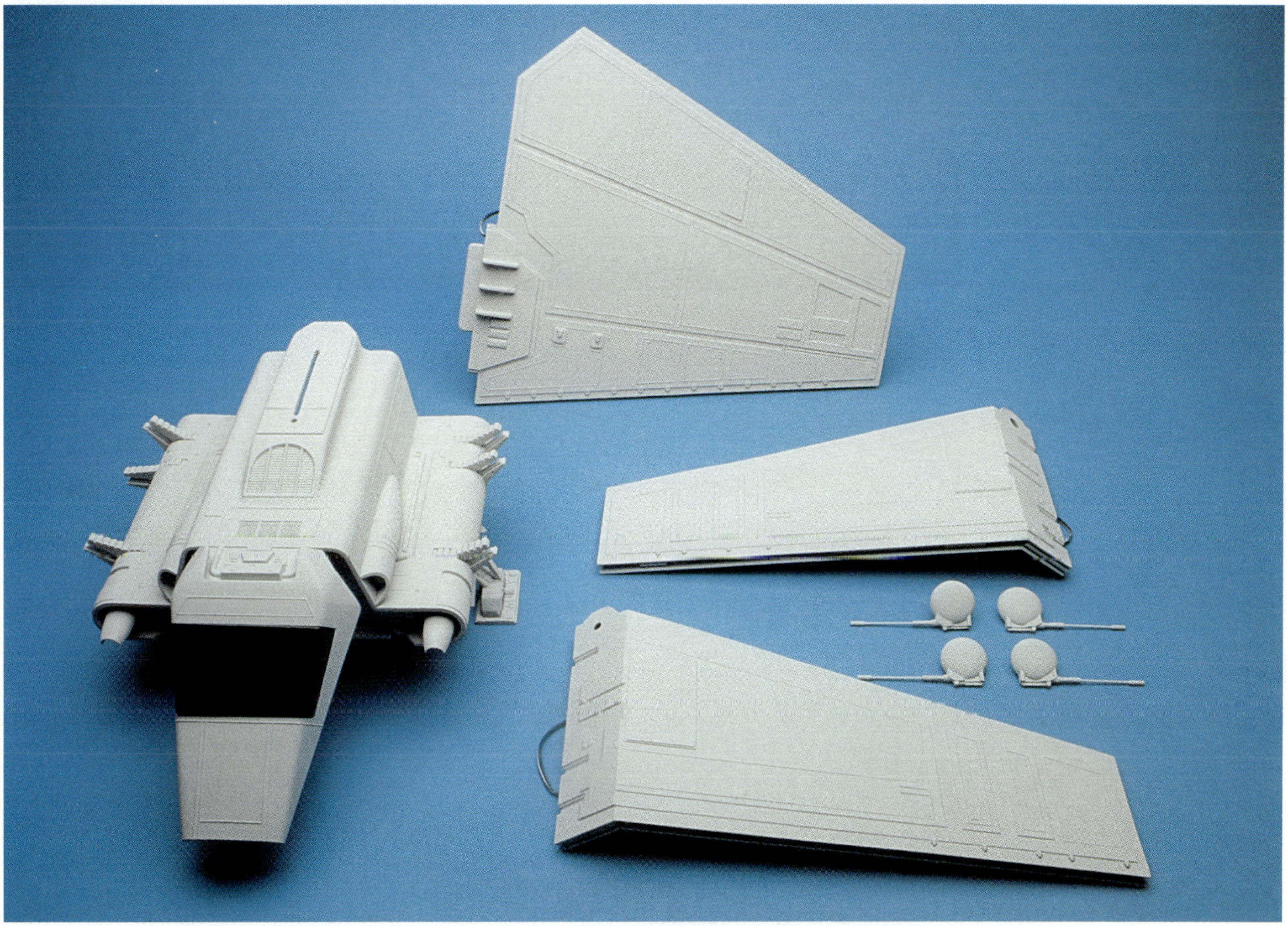

21 I gave all of the finished subassemblies an overall coat of Testors Model Master Camouflage Gray (no. 1733), applied with an airbrush. If you don't have access to an airbrush, this color is also available in spray cans (no. 1933).

22 There's a large D-shaped vent in the upper surface of the fuselage that should have a sooty black look. I masked the rest of the model and dusted it with flat black paint using an airbrush until I had the result I wanted.

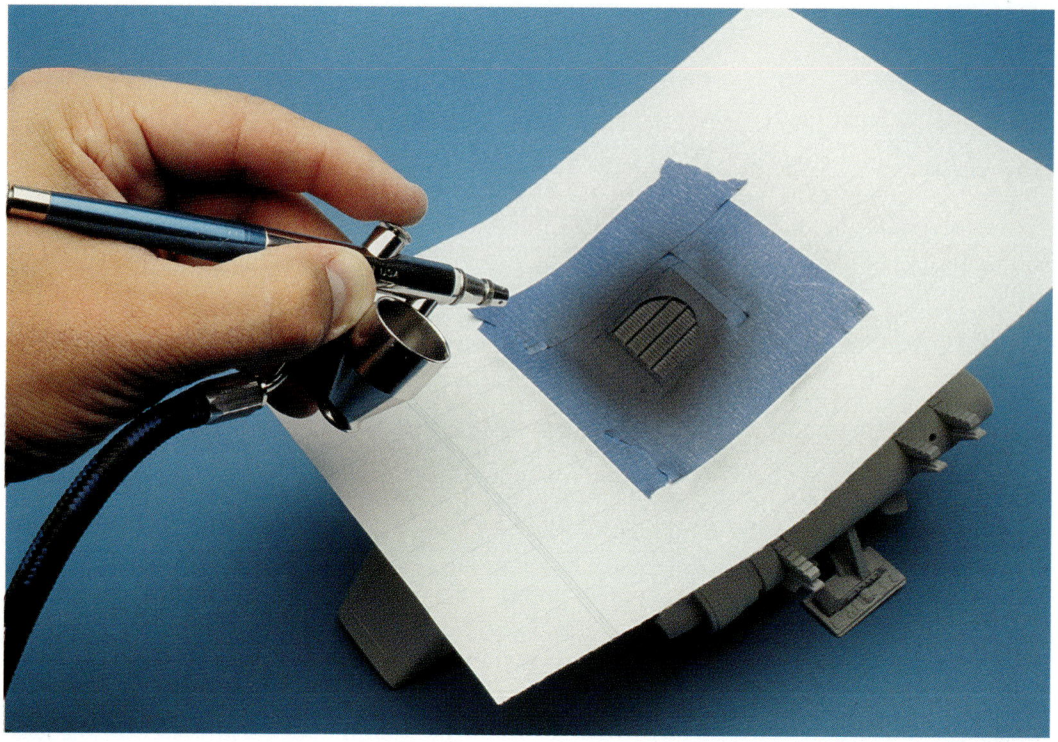

23 Several panels on the wings and fin should be painted light tan. Photos of the studio models in *From Star Wars to Indiana Jones: The Best of the Lucasfilm Archives* are a good guide.

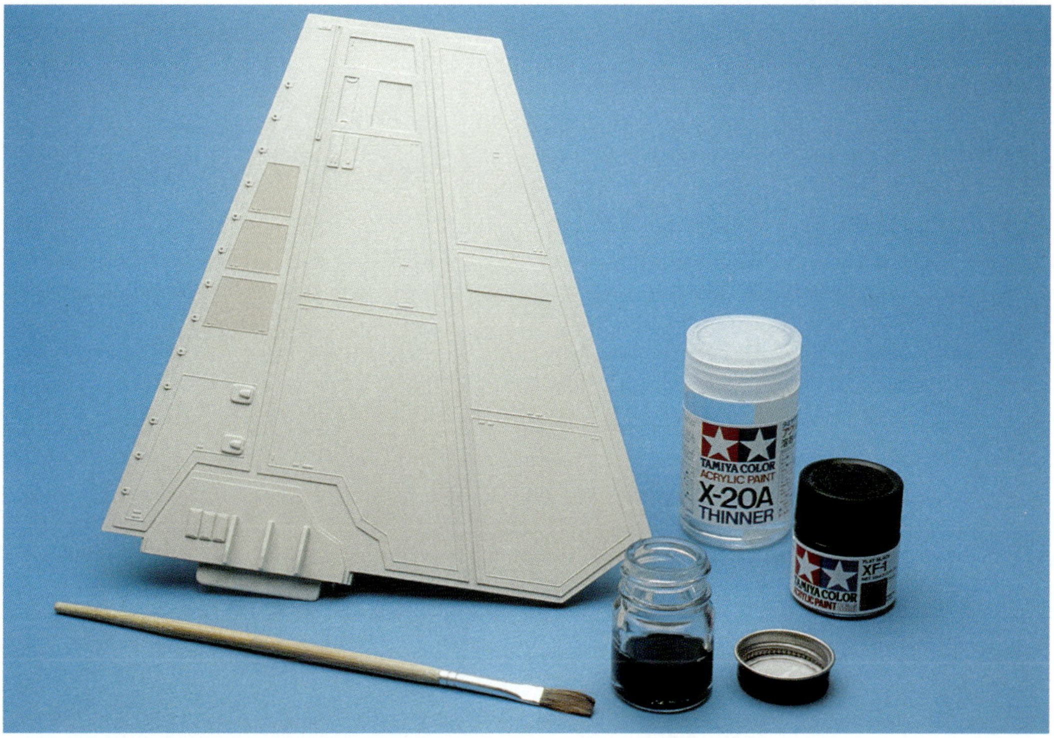

24 Unlike much of the Empire's hardware, the shuttles actually show a little wear and tear. Before assembling the model, I gave it an overall flat black wash, applying it with a wide brush.

25 I added the wings and tail to the fuselage and connected all of the formation-light wires to a single two-conductor wire that exits the model through a hole drilled inside one of the landing gear bays. Finally, I added the canopy and laser cannons, and did a little pastel weathering here and there, using the boxtop photo as a guide.

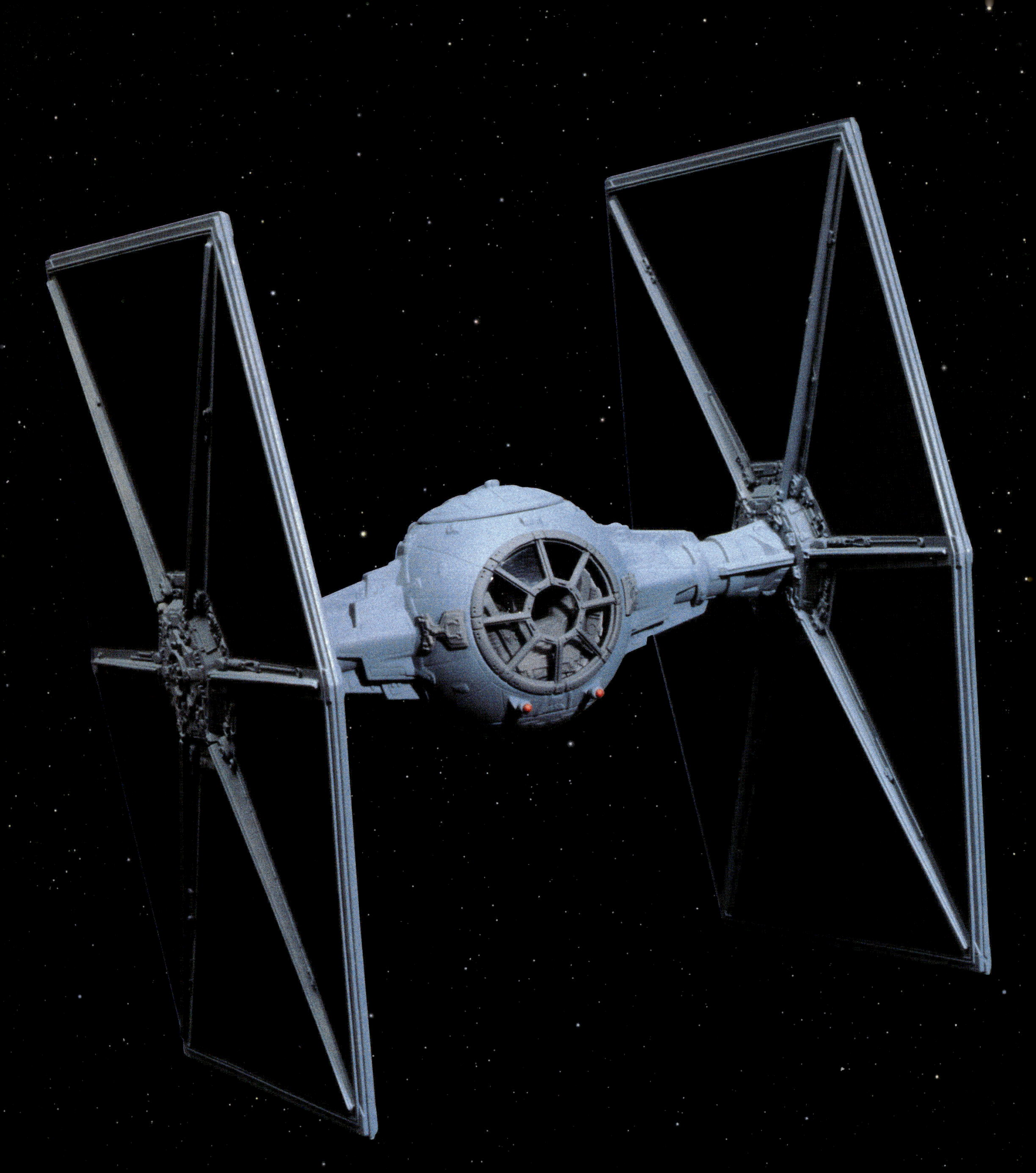

THREE

Imperial TIE Fighter

Star Wars fans like me have been adding lights to their models one way or another ever since the first kits appeared in the late 1970s. Some of the ships are easier to light than others; larger models like the Millennium Falcon and Imperial Star Destroyer have cavernous hulls and are easy to pile full of lights, wiring, and batteries.

The smaller models, however, require a slightly different game plan. When AMT/Ertl's Imperial TIE Fighter kit (no. 8438) appeared a couple of years ago, I wanted to light its twin engines, but the model's tiny interior (made even smaller by the full cockpit) presented the biggest problem. I knew that I'd have to add the smallest, most flexible lighting system I could come up with, then find some way to power it from the outside. I also wanted to detail and correct a couple of other areas on the kit. So, armed with a pile of parts catalogs and reference books, I went to work.

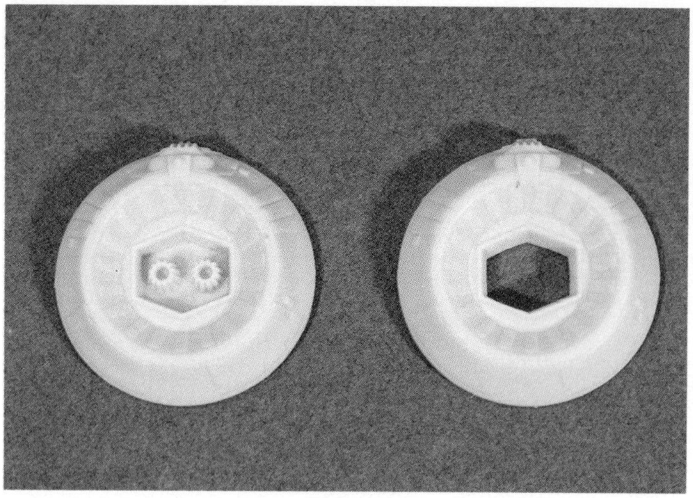

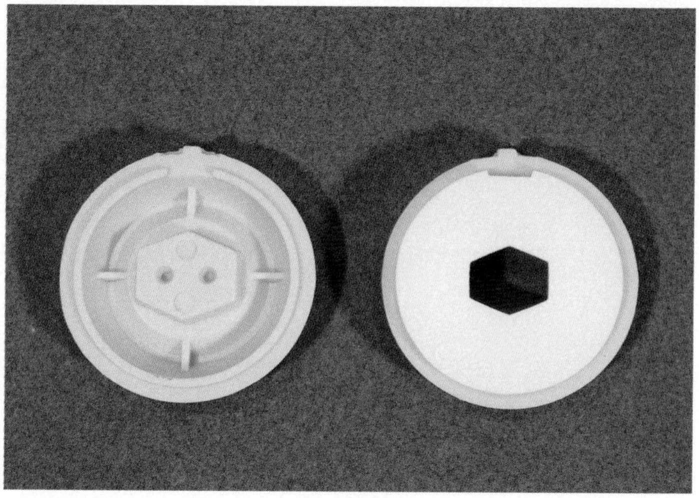

1 Looking at a videotape of the *Star Wars* Death Star trench sequence in slow-motion revealed a hexagonal window directly behind the pilot's head (and perhaps that I have too much free time). The kit's rear bulkhead (part 14), however, has two engine nozzles mounted in the space that should be the window, so I had some minor surgery to perform. I started by removing the nozzles and cutting open the window using a steel cutter chucked in my motor tool. A square file cleaned up the opening's rough edges.

2 The inside surface of the kit bulkhead isn't detailed, so I made a new interior wall from .030"-thick sheet styrene. After painting, I could trap a .010"-thick clear styrene window between the parts.

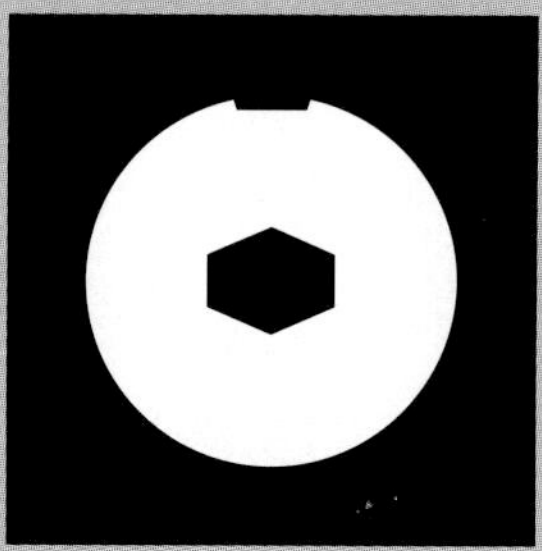

Interior Wall Template

3 I assembled the four-piece pilot figure using gap-filling super glue and attached him to a piece of styrene rod to make it easier to hold while I painted.

4 Although the pilot figure's uniform is black, by varying the shades and finishes I was able to break up the monotonous scheme. First I gave the figure a coat of Testor Aircraft Interior Black (no. 2040). After it had dried for a day or two, I hand-painted the helmet, hoses, gloves, boots, and belt with Testor Gloss Black (no. 1747) using a fine brush. The flat, slightly gray color of the suit contrasts nicely with the jet-black high-gloss details.

5 I had to modify the kit's cockpit slightly to accommodate the new window. I omitted the rear wall (part 20) and removed the rear-wall alignment tabs from the cockpit sidewalls (parts 6 and 7) and floor (part 5). Otherwise, I assembled the cockpit according to the kit instructions and airbrushed it Testor Gunship Gray (no. 1723). It's the same color I used for the interior of the fuselage, the front-window framing, and parts 15 and 16. I added the pilot to the cockpit and set the assembly aside.

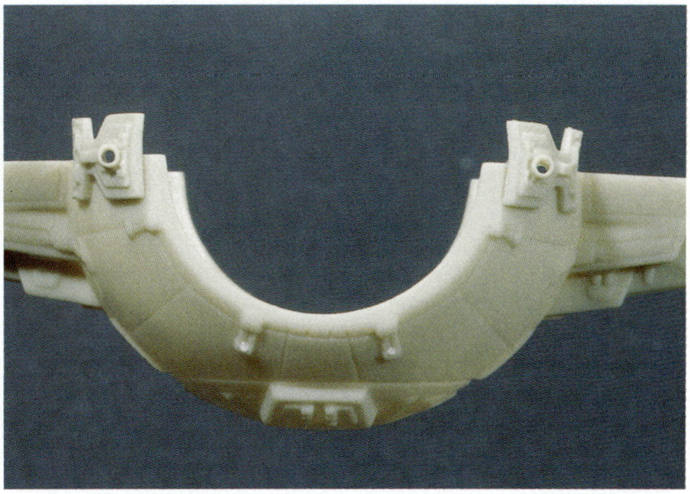

6 The TIE's engine nozzles are included on part 16. They're tiny, and I had to drill them out before I could light them from inside the model. They would be impossible to hold while I drilled, so first I glued them in place on the lower half of the fuselage (part 2) using gap-filling super glue, then I drilled the holes using a pin vise and a drill bit.

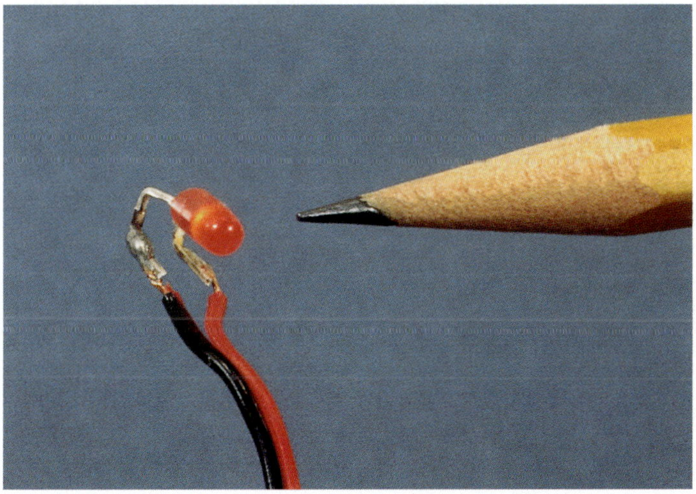

7 After determining the space I had inside the fuselage with the cockpit in place, I selected T-1-size red light-emitting diodes (LEDs) to illuminate the model; although they'd be small enough to fit, they'd still deliver plenty of light and would be easy to power with two AA batteries. I soldered the LEDs to two-conductor wire I found in the model railroad section of my local hobby shop. I soldered the red wire to each LED's positive lead (the longer of the two) and the black wire to the negative lead. Afterward, I bent the leads forward using a pair of needle-nose pliers to keep things as compact as possible.

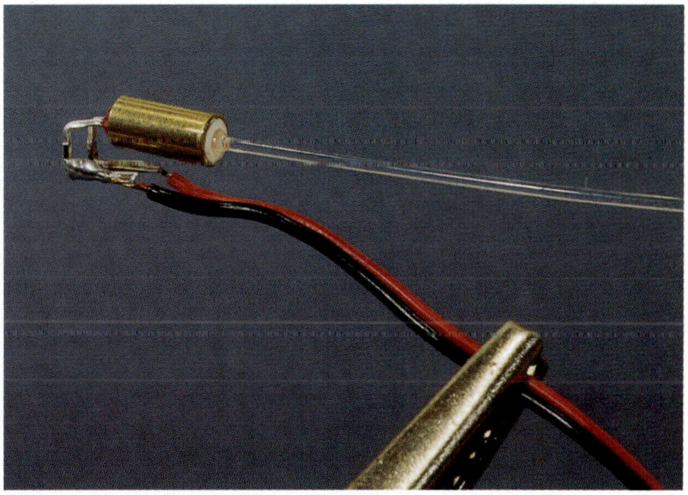

8 I decided that .040″-diameter fiber-optic strands from Plastruct (no. FOP-40) would be the best way to route the LEDs' light to the engine openings. The strands would fit inside the drilled-out engine nozzles and would still be flexible enough to bend around the cockpit sidewalls inside the model. To hold the end of the fiber-optic strand directly in front of the LED where it would transmit the most light, I slipped a short piece of plastic tubing inside a length of brass tubing, then epoxied a length of fiber optic in the end using five-minute epoxy. The finished assembly would be just the right size; it would fit over the end of the LED without taking up too much space.

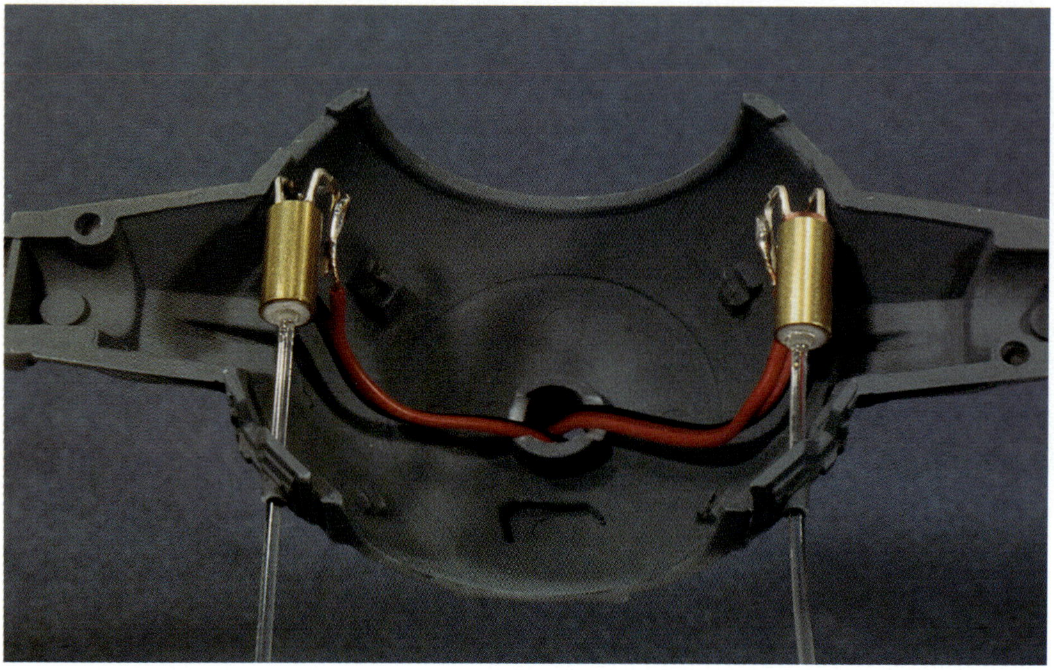

9 Test-fitting showed that both lighting assemblies would fit with room to spare. I also could pass both wiring leads through the stand hole in the bottom of the fuselage after I cut slots in each side of it using a steel cutter.

Combining the cockpit and the lighting system inside the fuselage is the trickiest part of the project. I trimmed the fiber-optic strands and used a three-grit polishing stick to shine up the cut ends. I passed them through the holes in the fuselage until $^{1}/_{32}$" extended outside the model, then glued them in place. After the epoxy cured, I glued the LEDs in place in their brass-tube holders and routed their wires through the stand hole.

Shoehorning the cockpit between the two lights and over all the wiring takes careful fitting and patience. Carefully pull the lights apart and slide in the cockpit. The fiber-optic strands should pass between the inner cockpit wall and the outer fuselage wall—it's a really tight fit, but it can be done. I cemented the cockpit in place, then added the upper fuselage half.

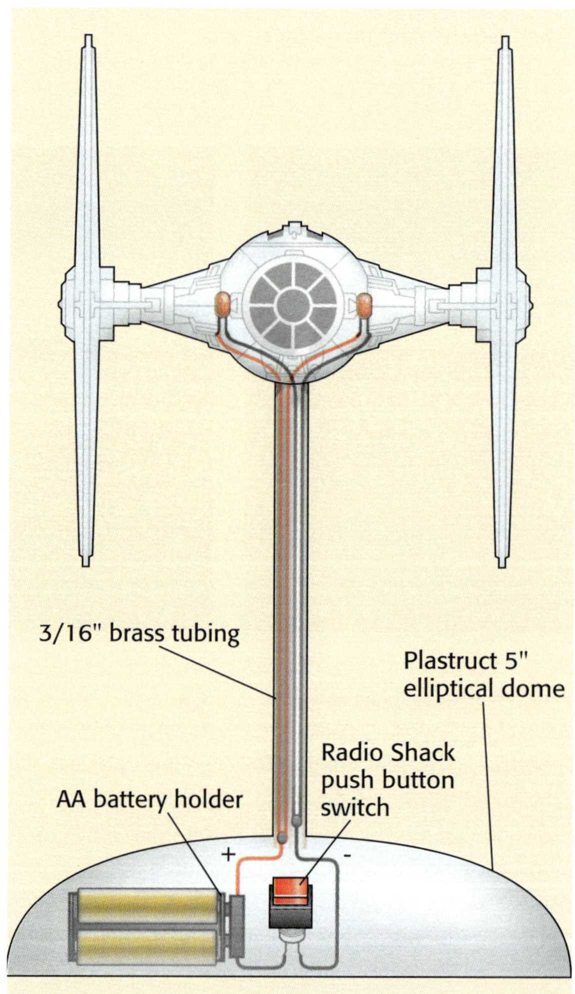

10 Determining the correct colors for the TIE Fighter was tricky. The main fuselage color varies from film to film, ranging from light gray to powder blue. A photo of one of the studio models I found in *From Star Wars to Indiana Jones: The Best of the Lucasfilm Archives* served as my main reference. I settled on Testor Intermediate Blue (no. 1720) for the main fuselage color. I airbrushed it on the fuselage after masking the window openings, then sprayed the framework on the hexagonal "wings."

11 When the Intermediate Blue paint was dry, I masked the wings' black panels and airbrushed them flat black. Believe it or not, painting the panels was the most time-consuming part of the project—it's easy to forget there are 24 of them until you have to mask each one!

After the paint had dried for a couple of days, I glued the wing panels in place on the main fuselage with gap-filling super glue, making sure they were properly aligned. I also installed the upper hatch, the front viewport, the modified rear bulkhead, and their clear parts using white glue.

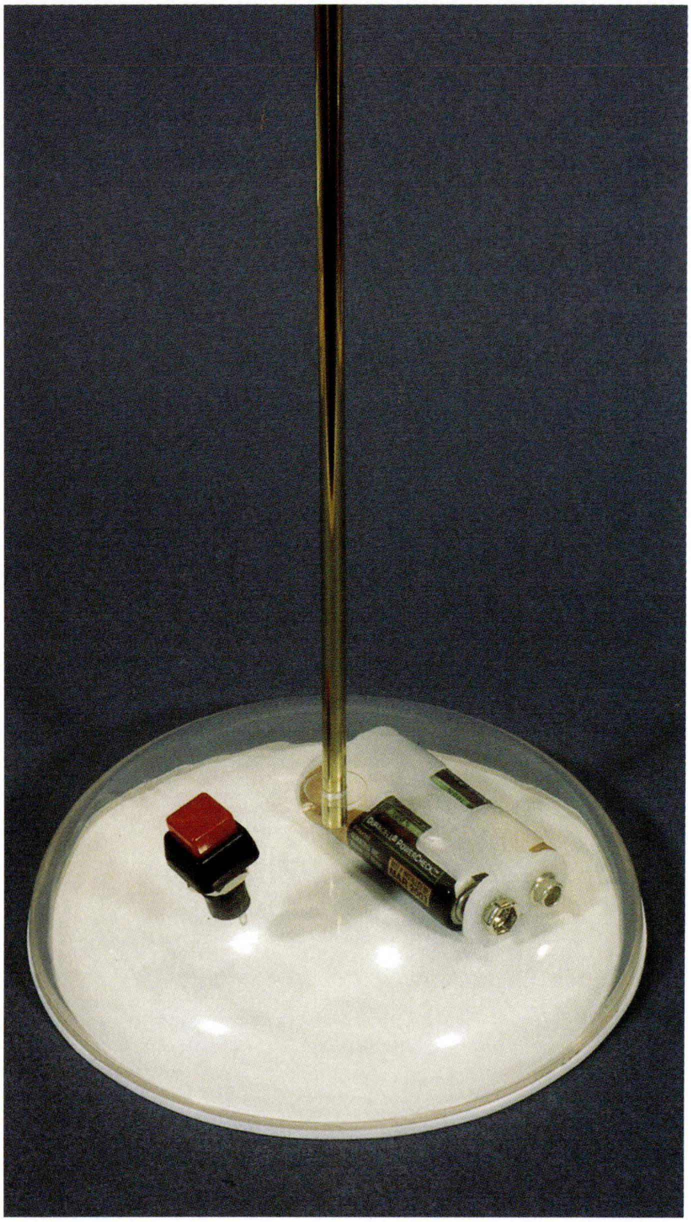

12 Although the kit includes a nice Death Star base, I wanted something a little more compact. I also needed to hide a pair of batteries and a bunch of wiring. The Plastruct catalog had what I was looking for; I ordered a 5″ clear acrylic elliptical dome (no. VHE-500). It's wide enough to support the model and big enough inside to hide all the hardware. A length of ¹¹⁄₆₄″-diameter brass tubing holds the model and hides the wires running from the battery. A red push-button power switch turns things off and on, and a circle of .080″-thick sheet styrene holds everything inside.

13 I painted the finished base flat black, but you could easily have skipped this step by ordering a black plastic dome (no. VHE-500BK) instead of a clear one.

14 The red push button lets viewers know there's more to the model than first meets the eye. Although the project's size made some of the work a little more difficult than usual, seeing the finished model with its tiny engines blazing away more than makes up for the trouble.

FOUR

Imperial AT-AT

The giant Imperial AT-ATs (All-Terrain Armored Transports) first lumbered onto the screen in 1980's *The Empire Strikes Back,* and can easily be counted among the film's most imaginative vehicles.

The AT-AT kit was originally released by MPC (kit nos. 1918 and 1929) and later reissued under the MPC/Ertl label (kit no. 8918). Although it hasn't been in production for a few years now, it's still relatively easy to find in well-stocked hobby shops and at swap meets. Best of all, the kit goes together pretty easily and offers plenty of opportunities to add detail.

During the battle sequence on Hoth, the Imperial walkers emit a sinister red glow from their viewport "eyes." I figured I could install a single, large light-emitting diode (LED) inside the head and hide a couple of batteries inside the model's large main hull. The wiring and kit modifications are straightforward. This would be a good first project if you've been thinking about adding lights to your models but have been uncertain about how to get started.

Also, unlike most of the Empire's hardware, the AT-ATs show signs of wear and tear. We'll add a little weathering to give the model an oily, battle-scarred look.

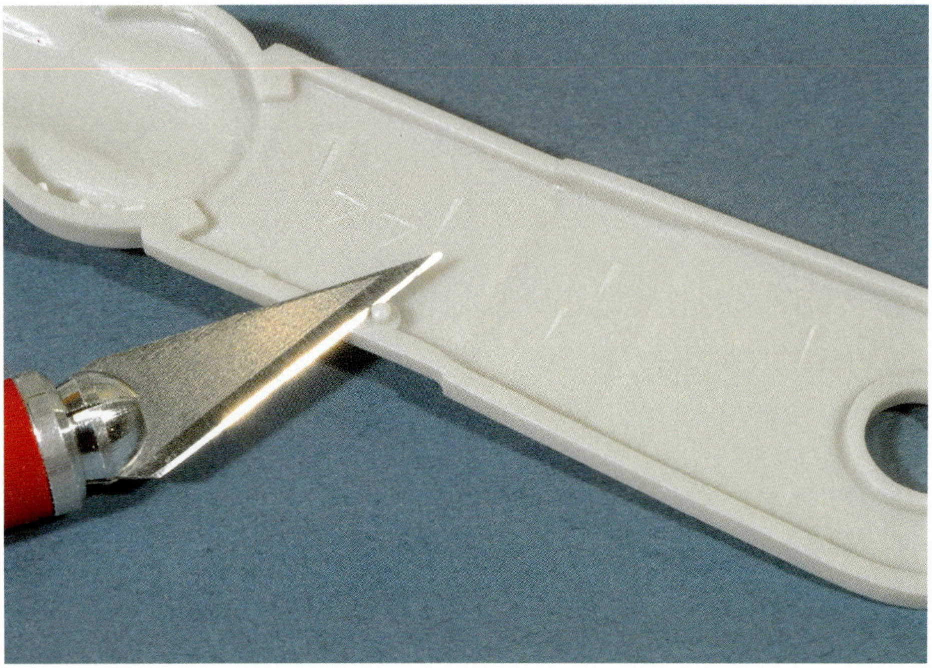

1 I started construction with the walker's legs. Many of the parts don't fit together too well right out of the box; cutting off the alignment pins with a hobby knife makes things easier to line up. Always test-fit the parts before gluing them together, especially if you use super glue, as I usually do.

2 A medium-grit sanding stick will remove all the leg seams after the glue dries.

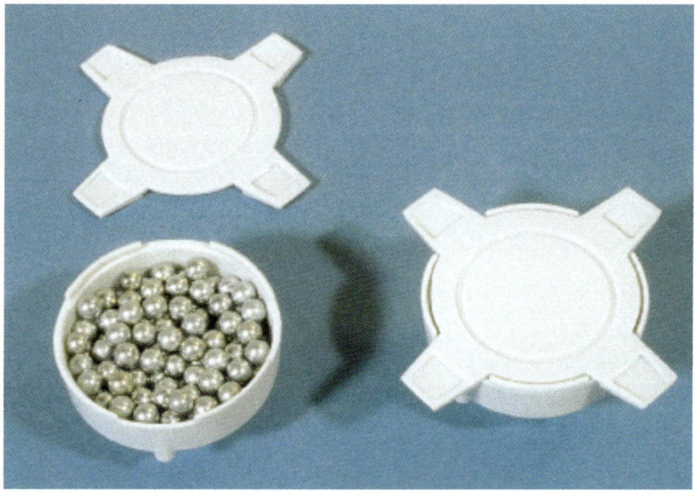

3 Not surprisingly, the walker is a little top-heavy. The first time I built one I had trouble keeping it on its feet, and I knew that adding a couple of batteries inside the hull would only make things worse. Luckily, the model's footpads are large and hollow, so I filled all four of them with BBs to add weight to the bottom.

4 I used a pencil-type soldering iron and rosin-core solder to hook up all the model's wiring. An "extra hands" parts holder makes it easy to align everything properly before you begin to solder. The LED's "bulb" is plastic, so to keep the soldering iron's heat from damaging it, I added an alligator-clip between the solder joint and the bulb to draw away the excess heat. Note that LED leads are of unequal lengths—the longer one is the positive lead. If the leads are reversed the LED won't light, so to keep things straight I used red wire for the positive side and black for the negative. Use plenty of wire—it's better to have too much than not enough, and any extra will be hidden inside the model.

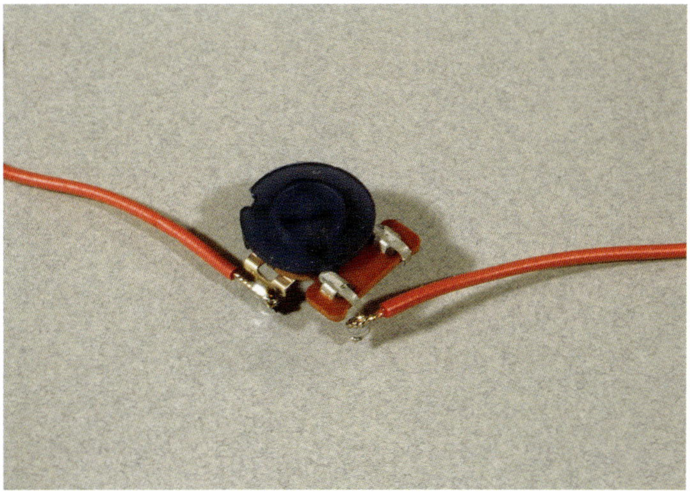

5 The two AA batteries I wanted to use are too powerful for the single LED, so I needed to dial down their voltage slightly. A rotary potentiometer (sometimes called a "pot") works like a tiny dimmer switch. Installing it in the circuit reduces the voltage to an acceptable level.

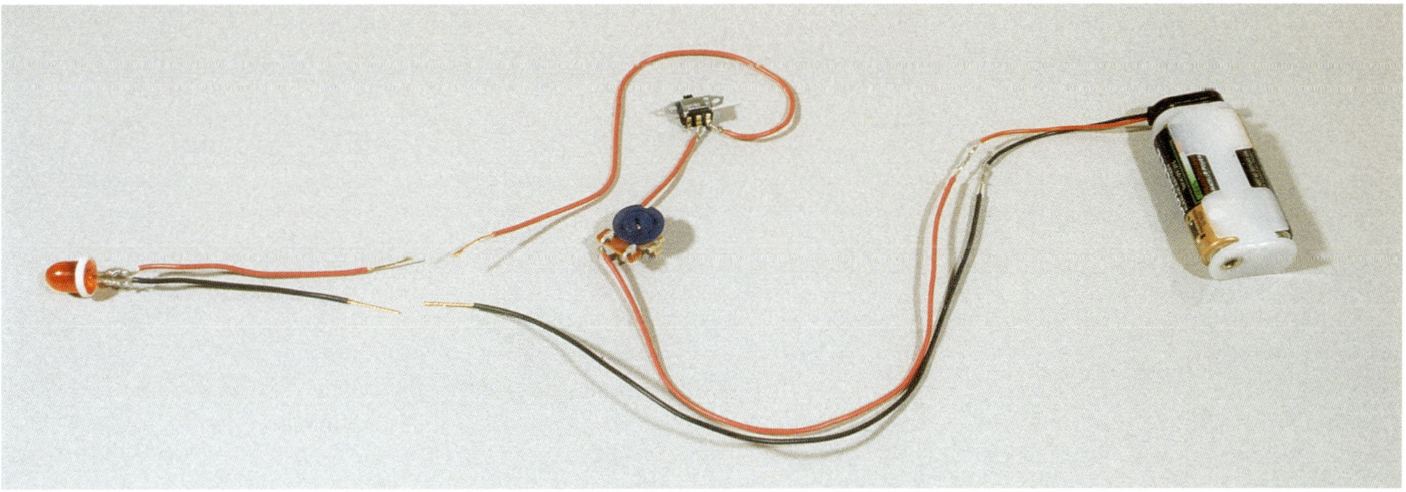

6 Here's the almost-finished lighting system, ready for installation in the model. Notice that the LED isn't wired to the rest of the circuit yet. This happens after it (and its wires) are installed in the model's head, and the batteries and switch are installed in the main hull.

7 The model's viewport is molded closed, so I drilled several holes inside the visor using a pin vise and a ¹⁄₁₆"-diameter drill bit.

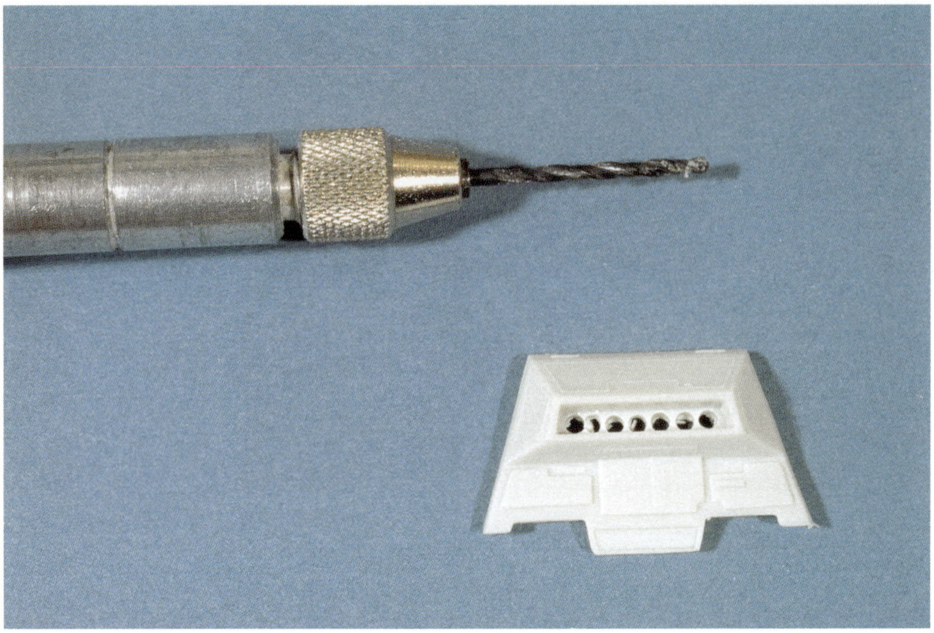

8 I used a hobby knife and a flat file to open the slot the rest of the way and smooth its edges.

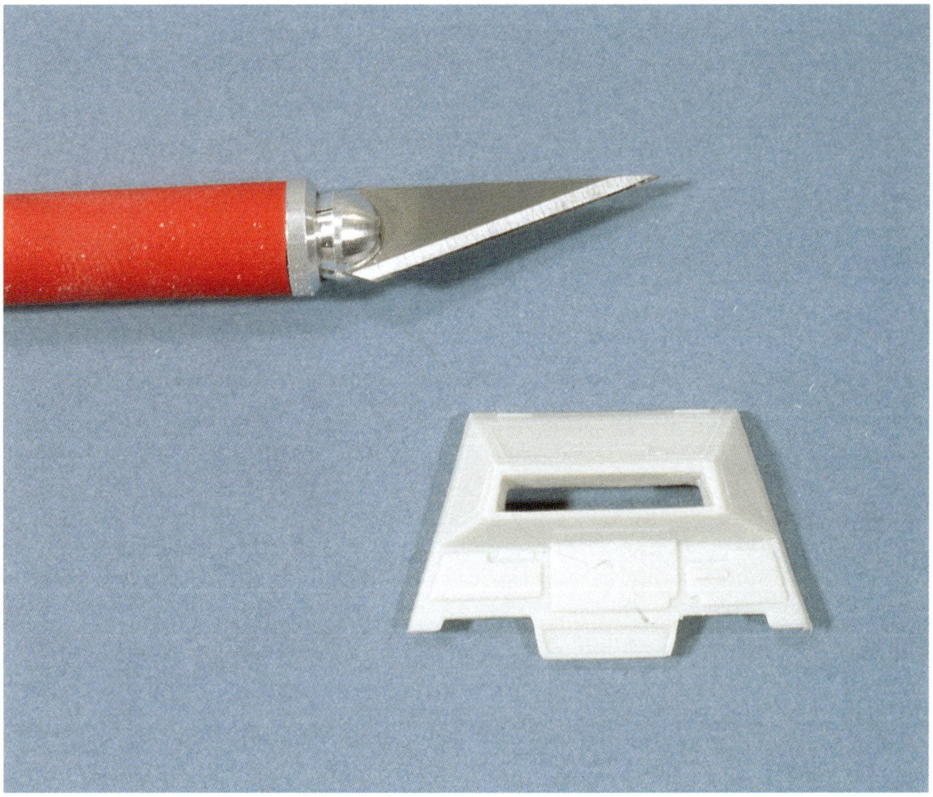

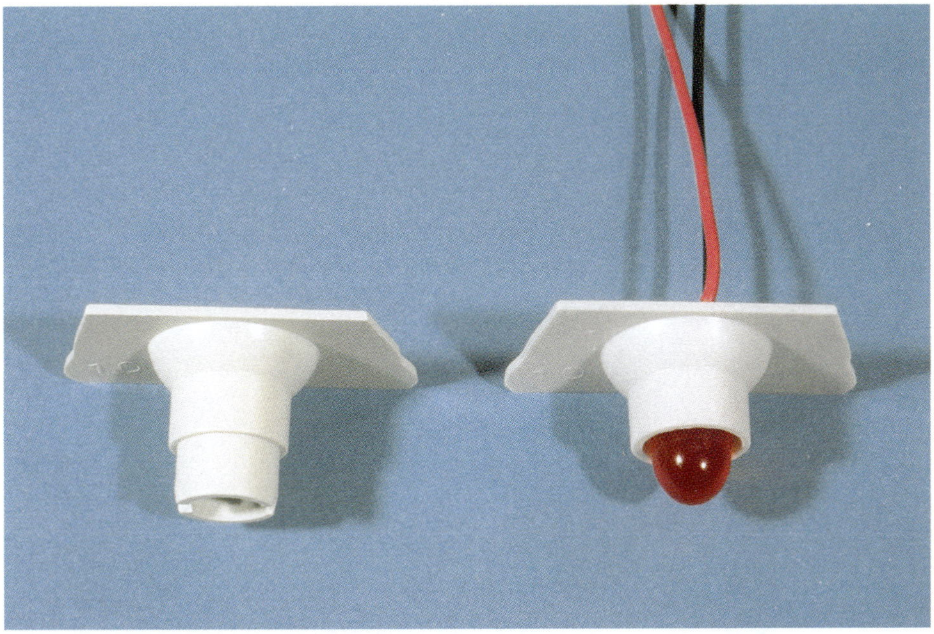

9 The tubular section of part 10 makes a perfect holder for the LED after a section is removed with a razor saw. The LED installs from the back, and the wires pass through the neck into the main hull, where the batteries are hidden. Although the kit provides parts to make the head poseable, they get in the way of the wiring, so I glued the head and neck in place.

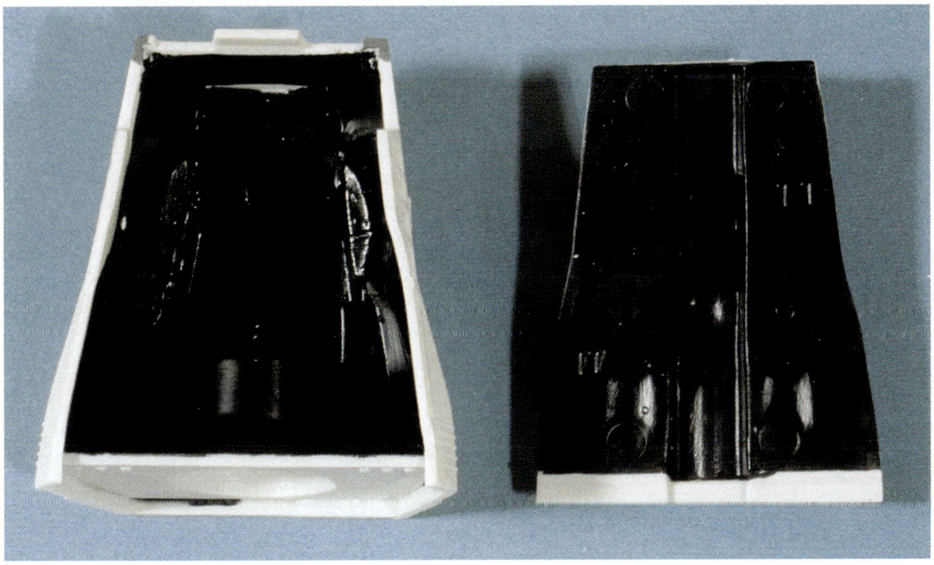

10 I assembled the head using gap-filling super glue, but left off the lower deck (part 11). To keep the LED's glow from showing through the head, I brushed a thick coat of black paint onto the inside surfaces. When it was dry, I brushed on a thick coat of silver paint. Afterward, I glued the lower deck in place.

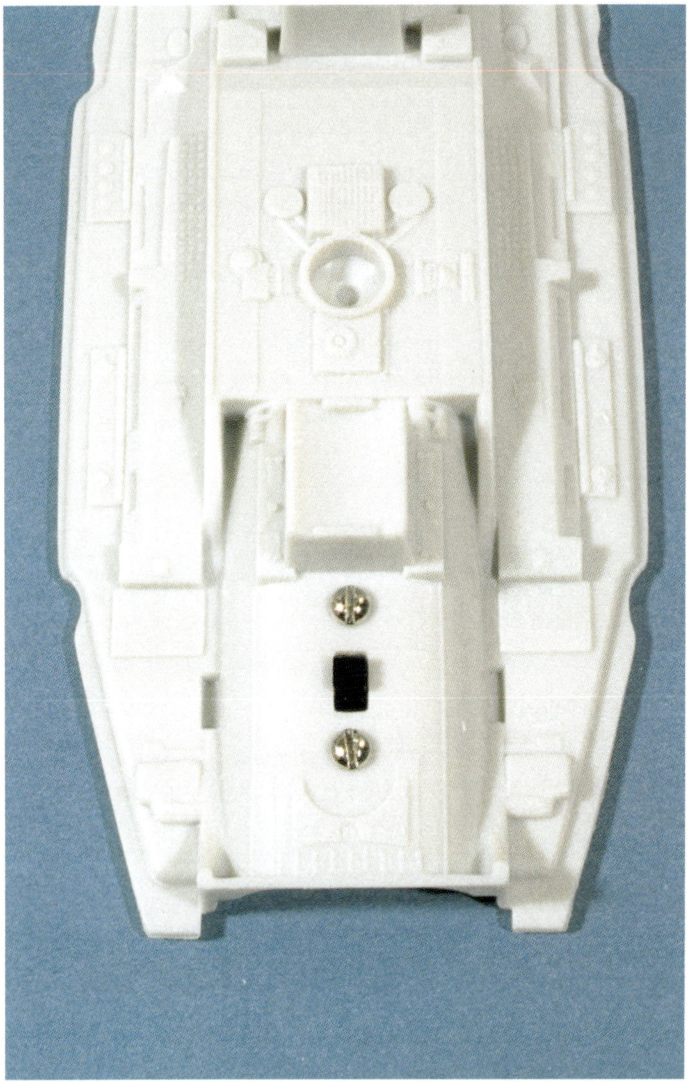

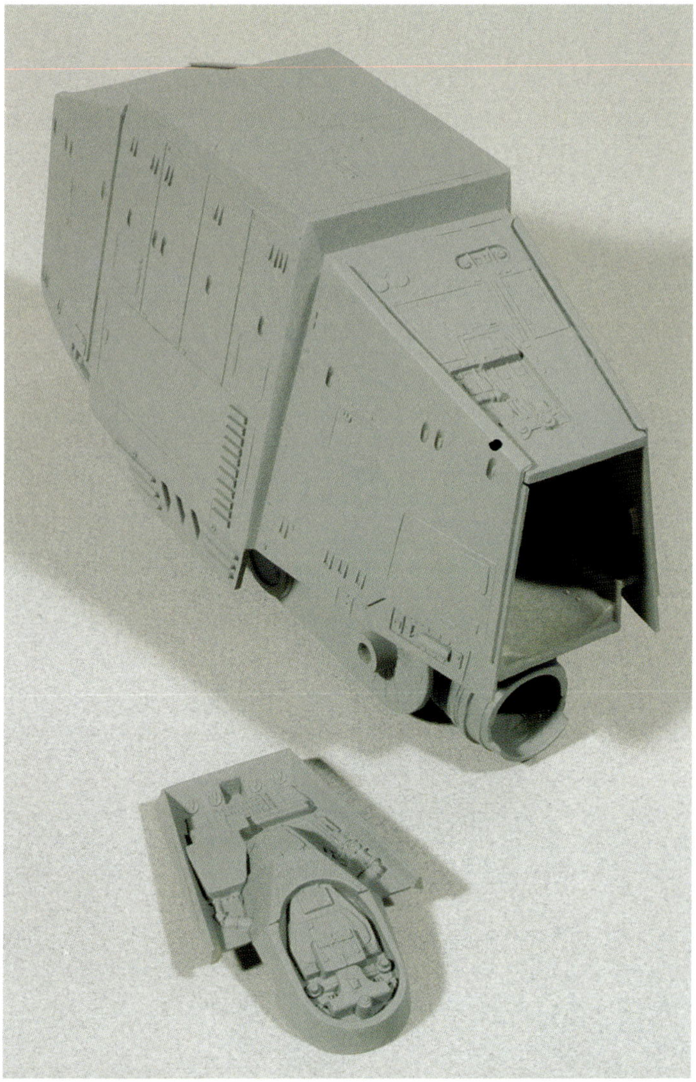

11 To hide the slide switch on the front end of the hull's bottom plate (part 21), I made the opening for it in the same way as I opened the visor. I drilled a hole, then shaped it into a rectangle using a hobby knife and a small file. Tiny nuts and bolts hold the switch in place.

12 Using gap-filling super glue again, I assembled the hull. The hull's rear panel (part 5) will stay in place without glue, making it easy to change the batteries inside when they wear out.

13 The hull has a few fit problems, particularly where the top panel (part 1) meets the sides. Before I filled and sanded the seam lines, I covered all the raised panel lines with masking tape so I wouldn't accidentally remove them.

14 I airbrushed the model's subassemblies with Testor Model Master Aggressor Gray (no. 1794); it seemed to be the best out-of-the-bottle match. After the paint had dried for several days, I started weathering the model. To simulate the model's oil stains I thinned a small amount of black acrylic paint until it was the consistency of black coffee. Using pictures of the production models as a guide, I added oil streaks with a fine paintbrush, letting the color build up slowly.

15 The stains should run straight down the sides of the model, gradually thinning to a point. When I was satisfied with the oil stains, I thinned the paint even more, until it was almost transparent. Using a wide brush, I gave the model an overall wash, letting the color flow into all of the recesses.

16 I glued the LED into the opening in the back of the head, then glued on the neck after routing the wires through it. I soldered the leads to the wires from the hull, covered the soldered joints with electrical tape, and glued the head assembly to the main hull. This is where all the extra wire I mentioned earlier comes in handy—the hot soldering iron never has to get too close to the delicate plastic model parts.

17 I added the legs to the hull using super glue, making sure all four feet were touching the ground. I installed the blaster cannons on the head, and the AT-AT was finished and ready for battle.

STAR WARS™
NABOO STARFIGHTER™

FIVE

Die-cast Naboo Fighter

When it first appeared in *The Phantom Menace,* the sleek Naboo Fighter was distinctively different from the *Star Wars* ships that had come before it. Unlike the battered, utilitarian ships of the first three films, the Naboo fighter looked more like a piece of art deco, with its streamlined design and bright yellow-and-chrome finish. It was a natural choice for AMT/Ertl's line of *Star Wars* kits.

Modelers were surprised to see two kits of the Naboo fighters on hobby shop shelves—a snap-together plastic version and a screw-together die-cast metal version. The fact that the model is die-cast makes sense when you remember that in addition to its plastic model line, AMT/Ertl is one of the world's largest manufacturers of die-cast metal models.

Most likely aimed at die-cast collectors who might not be avid model builders, the kit has painted parts and can be assembled into a good-looking model with a few simple tools. It's a good project for younger modelers, too. With just a little help they can have a finished model on their bookshelf in a couple of hours.

If you're more interested in the plastic version of the kit, don't worry. There's a sidebar at the end of this chapter that outlines building, painting, and detailing the plastic version. Let's get started.

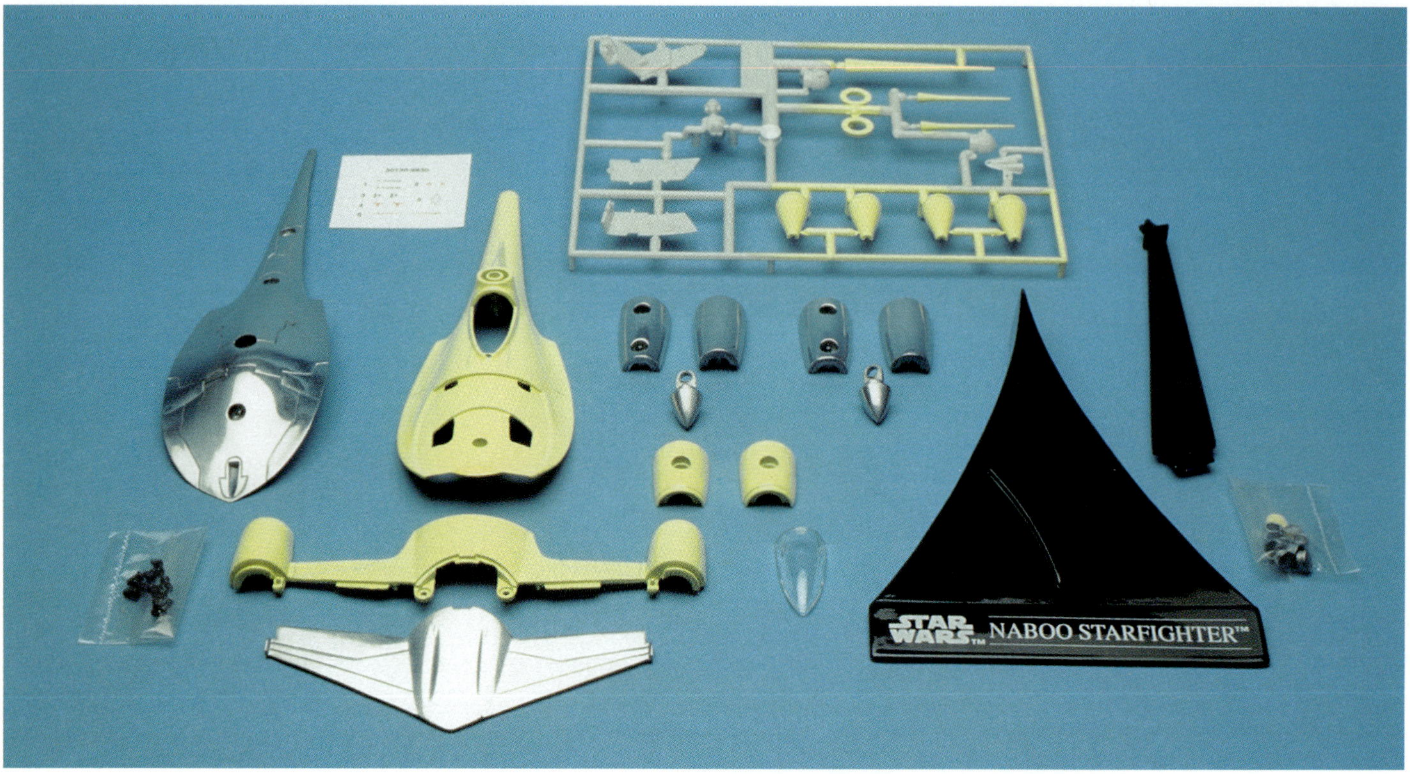

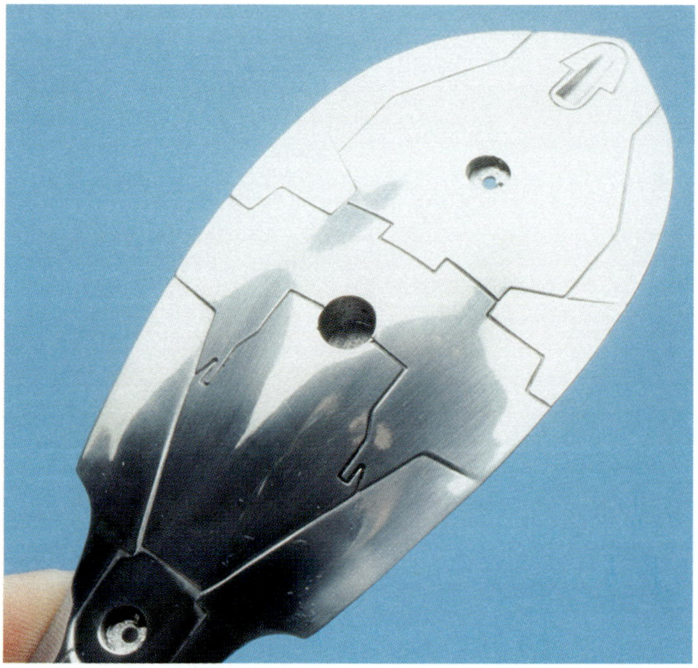

1 Here are the kit's components, straight from the box. Although some of the parts are unpainted, the upper fuselage parts, as well as some of the parts on the plastic tree, have a smooth coat of gloss yellow paint. Even the base is painted and features tampo-printed (pad-printed) lettering.

2 The kit's natural-metal parts arrive polished to a high shine. With a little work they could be installed in the plastic version of the kit, replacing its chrome-plated parts.

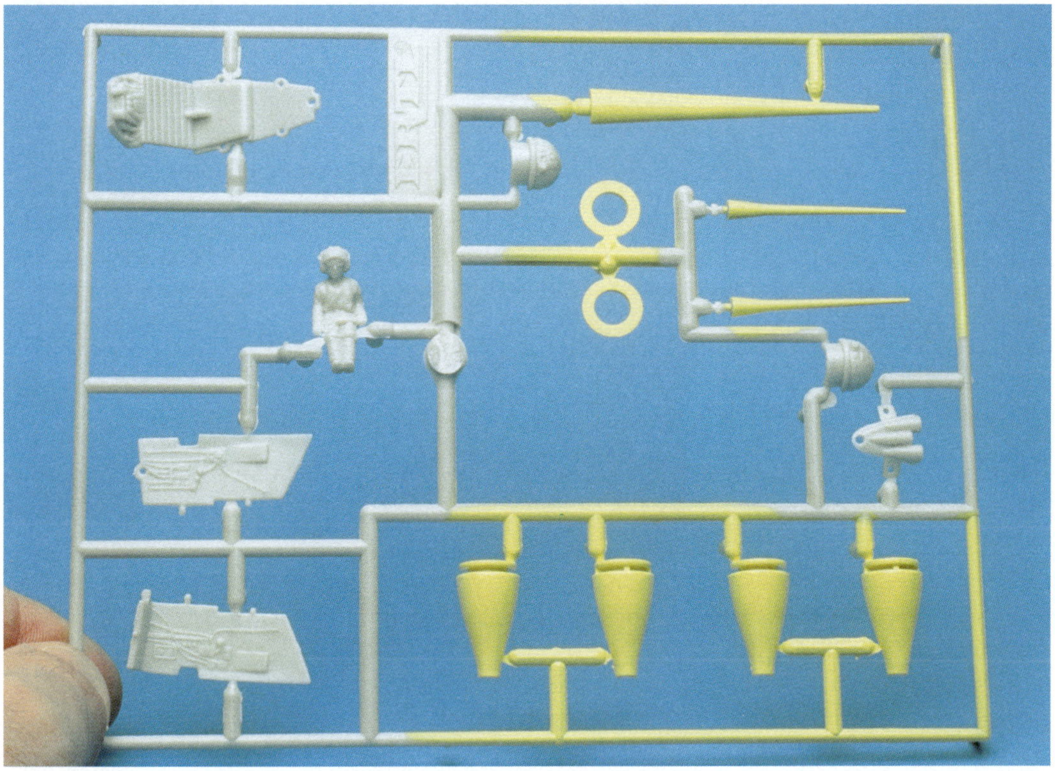

3 A single plastic parts tree is included in the kit, and sections of it have been painted gloss yellow. The cockpit and pilot need some work, though, so let's get started there.

4 A good pair of parts cutters makes it easy to remove the thin cockpit walls from the parts tree. Afterward, clean up the attachment points with a hobby knife and a medium-grit sanding stick.

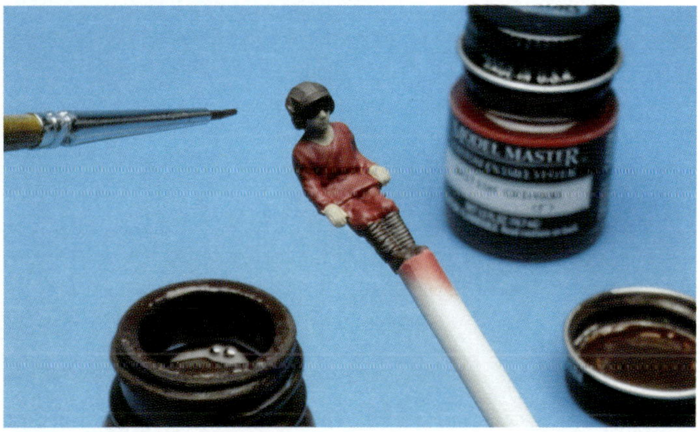

5 The kit doesn't have the world's most detailed pilot, but a little careful painting with a detail brush improves things considerably. I attached the inch-long figure to a straight section of a leftover parts tree with a small dot of super glue. The tree makes the pilot easier to hold while you paint, and when you're finished, the figure can be popped loose.

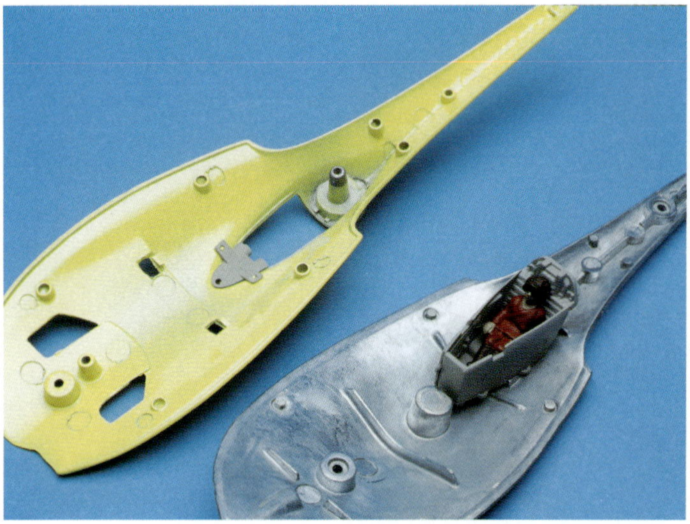

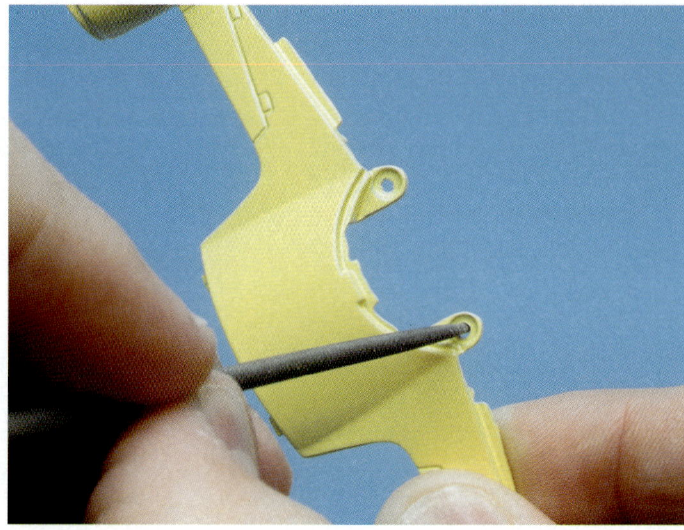

6 I painted the assembled cockpit and instrument panel medium gray, added the pilot, and glued them in place on the lower fuselage half. Gap-filling super glue or epoxy is great for joining dissimilar parts like the die-cast metal fuselage and the plastic cockpit parts. Regular model glue won't stick to the metal parts.

7 Some of the model's screw holes were molded closed; I used a round file to open them up.

8 Tighten the Phillips-head screws with a small screwdriver. Be careful not to overtighten them, or the kit's delicate molded-in threads will strip.

9 After the screws are in place and tightened, lock them in place with a drop of super glue. Capillary action will carry the glue in around the threads, and it will keep parts from loosening later on.

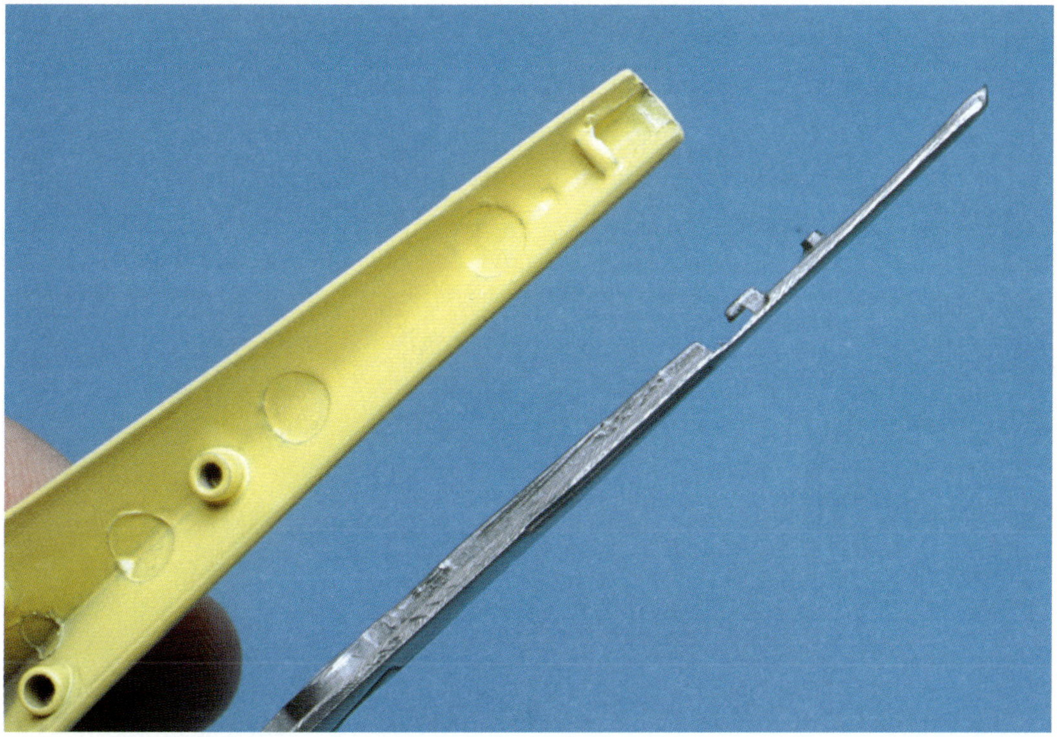

10 There's a hook molded into the lower fuselage half; make sure you hook it through the loop in the upper half when you assemble the fuselage—it's easy to miss.

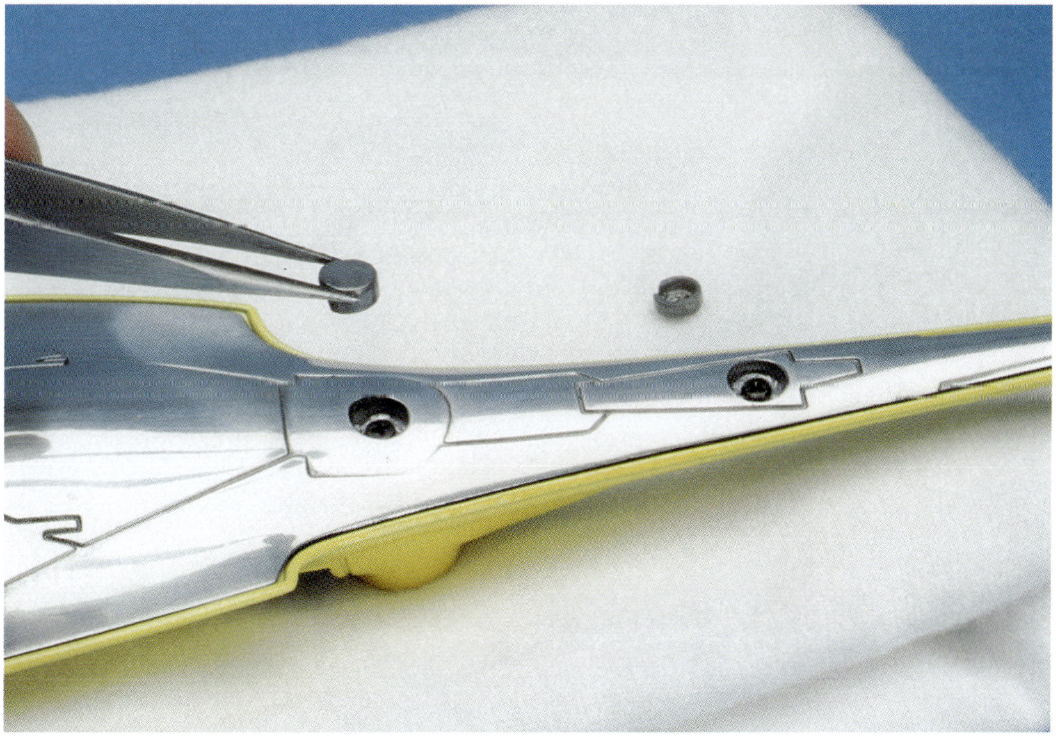

11 Three screws hold the upper and lower fuselage halves together. They're hidden under polished-metal caps under the model. Each of the caps is numbered inside; the instructions show where each one goes. A drop of super glue or a little epoxy will help keep the caps in place. I placed the model on a folded-up towel to protect the finish from scratches and chips while I worked.

12 To give the natural-metal parts a slightly different look, I applied Bare-Metal matte aluminum foil to the center section of each of the "wings." The recessed panel lines make it easy to trim the excess foil with a sharp hobby knife blade.

13 The plastic engine exhaust cones are trapped in place by the lower half of the engine nacelle. As with the fuselage, these screws are hidden under press-in caps.

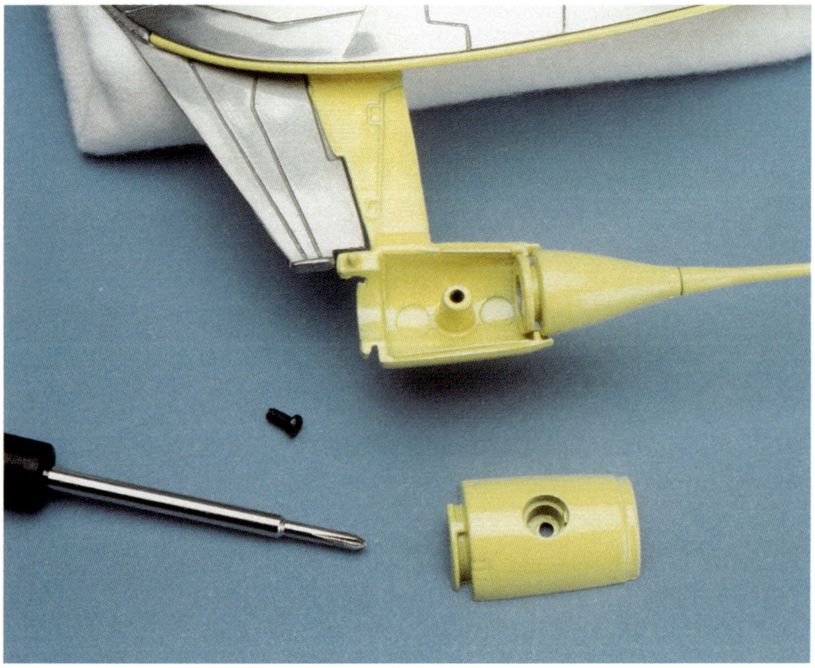

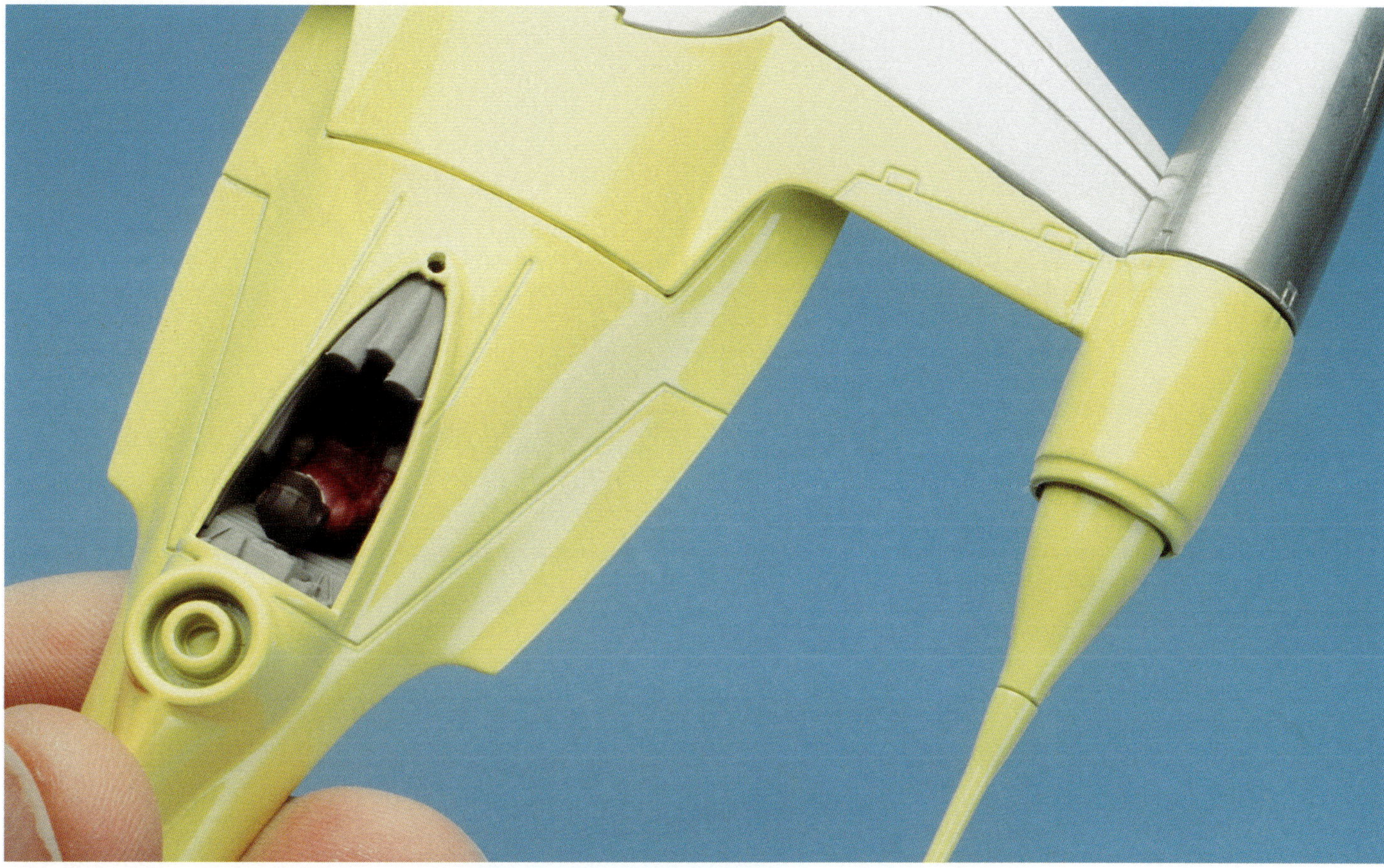

14 Here's the finished fuselage, ready for its canopy and R2 unit. A black wash in the recessed panel lines would give the model a more realistic look.

15 The canopy has a delicate frame that should be painted to match the rest of the model. If you don't feel like trying to match the paint yourself, Polly Scale Seaboard Air Line Yellow acrylic paint (no. 414384) is a pretty good straight-from-the-bottle match for the kit's factory-applied yellow paint. Although I masked and airbrushed the canopy, you could also use a detail brush (and a steady hand) to paint it.

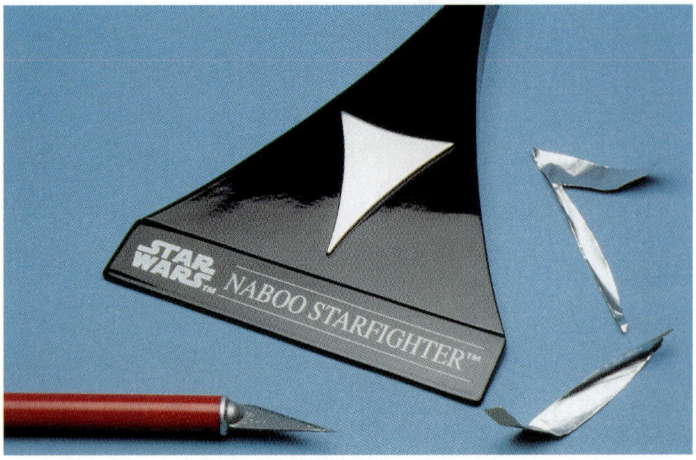

16 Ironically, with all the kit's metal parts, the R2-D2 head is assembled from two unpainted plastic parts. I substituted the chrome-plated version from the plastic version of the kit and detailed it with navy blue paint.

17 While I had out the Bare-Metal foil, I covered the raised triangular emblem on the die-cast display base.

18 With just a little extra work, the finished die-cast model is almost indistinguishable from its injection-molded plastic cousin.

BUILDING THE PLASTIC NABOO FIGHTER

If the die-cast version of the Naboo Fighter doesn't appeal to you and you're more interested in adding the plastic version to your collection, here's a quick review of how the standard kit goes together straight from the box.

One of the first three Episode I kits released by AMT/Ertl, the 1/48 scale Naboo Fighter kit (no. 30117) features snap-together construction. I was surprised to find a large chrome-plated sprue with the ship's forward wing, engine nacelles, and fuselage underside, and a two-piece R2-D2 head. The rest of the kit's 20 parts are molded in bright yellow and clear plastic, with markings provided on a tiny decal sheet. A two-piece display supports the model; no landing gear is provided.

I decided to break the rules a bit and glue my kit together, so I started by removing the snap lugs with a sprue cutter and a razor saw. Following the kit's four-step instruction sheet, I started assembly with the fuselage. I glued the rear half of the wing (part no. 2) to the upper half of the fuselage and installed the three-piece engine-exhaust cones. A little work with a sanding stick and some gap-filling super glue filled the seams between the various parts.

Just a little extra work will make the snap-together plastic Naboo Fighter kit into a good-looking model. It's the same scale as the Droid Fighters kit, too, which opens up plenty of diorama possibilities.

Although it's molded as part of the fuselage, the cockpit is adequately detailed and looks even better with the painted pilot in place. The canopy fits well; I glued mine in place with white glue.

I masked the canopy with self-adhesive chrome foil and airbrushed the fuselage Testor Insignia Yellow. While it was drying, I moved to the forward section of the wing and engine nacelles.

Although I was tempted to use the chrome-plated parts right out of the box, the plating would have made it difficult to fill the seams between the engine-nacelle parts. I soaked the parts in bleach to remove the plating, filled the seams with gap-filling super glue, and airbrushed them with SnJ Spray Metal. When they were dry, I added them to the fuselage with super glue. The kit decals are a little thick, but snuggled down after a dose of Micro Sol. A final coat of Testor Dullcote sealed them. A little weathering and a dark wash along the panel lines really give the model a realistic appearance.

I airbrushed the display base flat black, then removed the paint from the raised lettering with a medium-grit sanding stick. I added a little Bare-Metal foil to dress up the triangular emblem, then added the clear support.

My Naboo Fighter took only a few evenings to build, and even with its simple construction, is a good-looking replica of the film ship. *Star Wars* fans of all ages will be able to handle it easily.

SIX

Battle Droid

While spacecraft have always been the core of AMT/Ertl's *Star Wars* line, a few figure kits (both plastic and vinyl) have appeared from time to time. It wasn't much of a surprise when a plastic Battle Droid kit appeared with the release of *The Phantom Menace*—the tall, spindly droids appear in droves all through the movie, in a variety of roles.

The kit (no. 30124) includes a single Battle Droid and an STAP (Single Trooper Aerial Platform) for it to ride. A closer look at the figure reveals that it's articulated—all its arm, leg, and head joints are hinged to make the droid poseable. Originally AMT/Ertl had plans to release several versions of the Battle Droid kit in which the working features would be more appropriate, but although additional kits appeared in the company's catalog, none was ever released.

The original kit is still versatile, though, and after a little planning, I decided to put aside the kit's aerial platform parts and concentrate on building the Battle Droid as a standing figure, adding detail and an accurate paint scheme along the way.

1 The kit instructions start with the droid's head, so let's do the same. The "neck" supports (parts 5 and 6) in my kit had tough-to-fix sink marks in the plastic. Instead of trying to fill them, I hid them under panels of .010" styrene sheet.

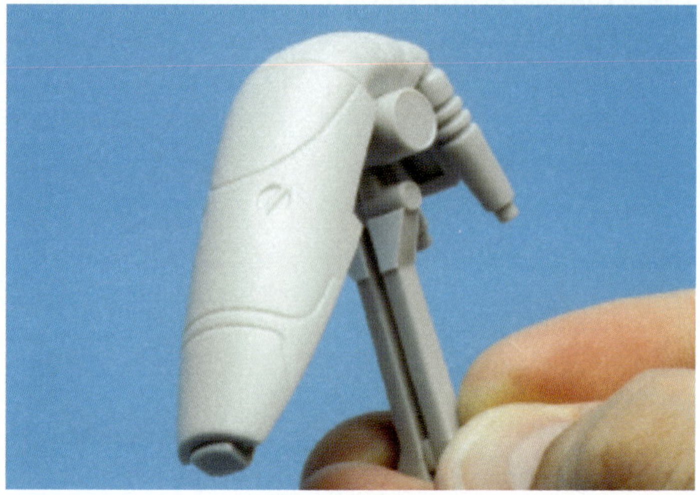

2 The styrene panels that cover the sink marks don't look all that out of place on the model, especially later after it gets a coat of paint.

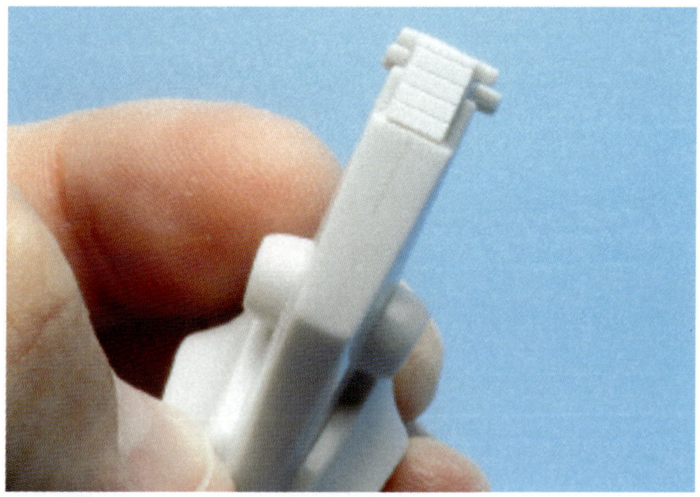

3 I used sheet styrene to correct several of the kit's problem areas. Here I've used grooved sheet to cover a problem area on the model's pelvis.

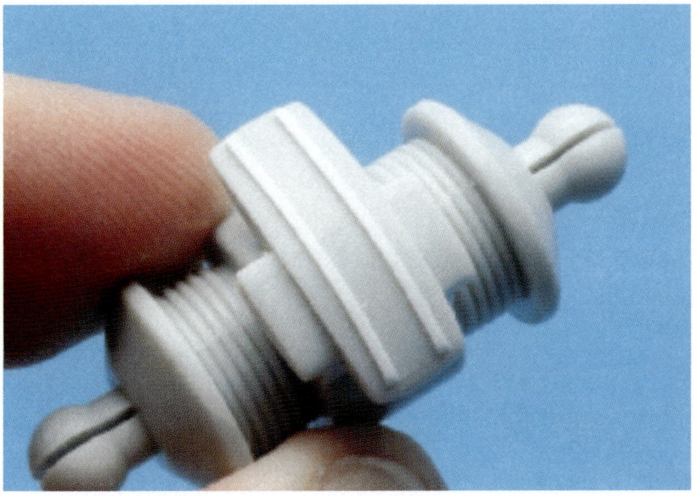

4 There are lots of styrene shapes available from companies like Evergreen and Plastruct, and all of them come in handy at one time or another. The molded-in ribs on this section of the model weren't very sharp, so I sanded them off and replaced them with sections of styrene strip, which I glued in place with super glue.

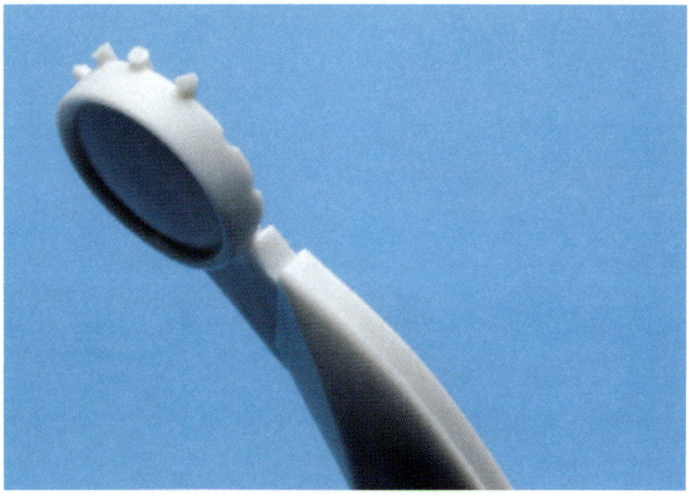

5 The model's leg and arm joints are hinged and have click-stop notches to hold the model in different poses. The working hinges were probably included for the "standing" versions of the kit that were announced but never produced. The notches are unrealistic, so I stretched some of the parts-tree sprue and glued short pieces into each of the half-round notches using gap-filling super glue. Afterward I trimmed the excess plastic away and sanded the joints smooth.

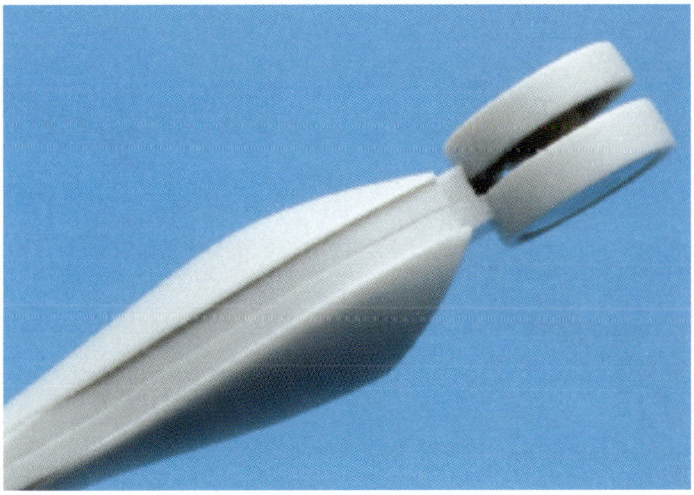

6 Test-fitting revealed a major problem with the model's legs. A prominent seam on the back of each leg would be tough to fill.

7 Sheet styrene solved the problem. This time I used a strip of .010″ sheet to cover each of the seams.

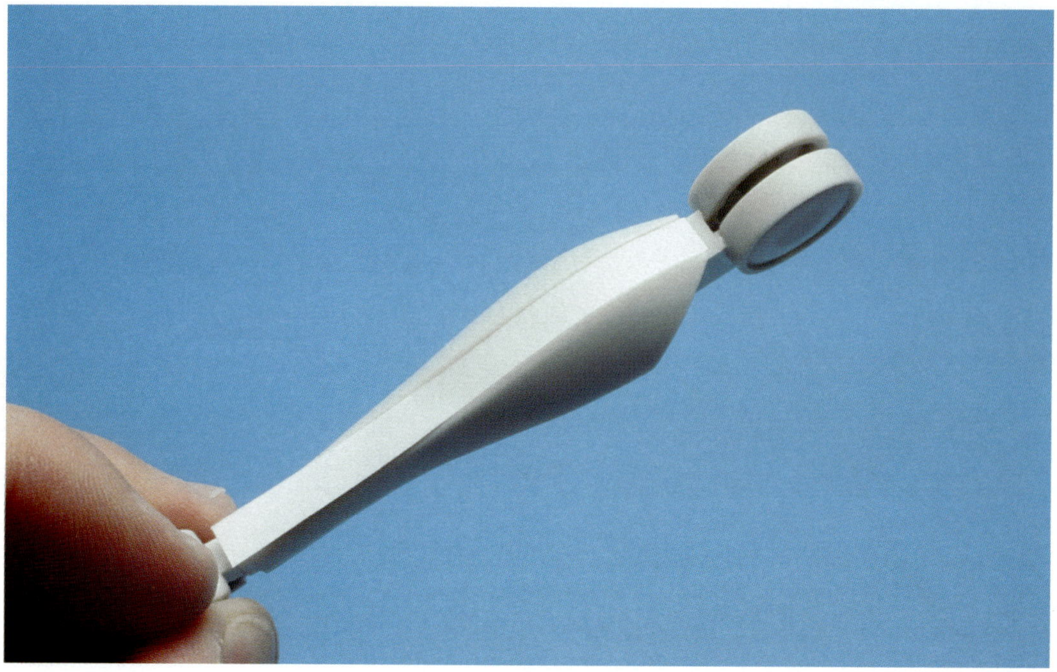

8 A little careful sanding stick work smoothed the styrene strip into the leg. Be particularly careful when you shape the curved lower section of the strip.

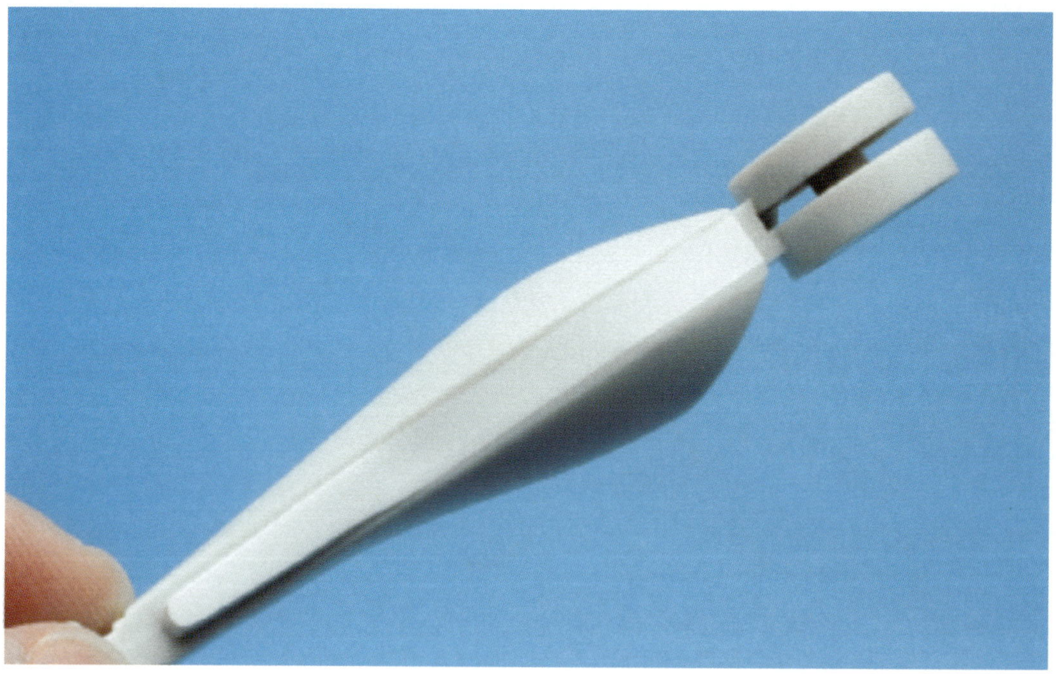

9 The fit on the front of the legs isn't all that great, either, but a little gap-filling super glue and some sanding stick work should help fix things.

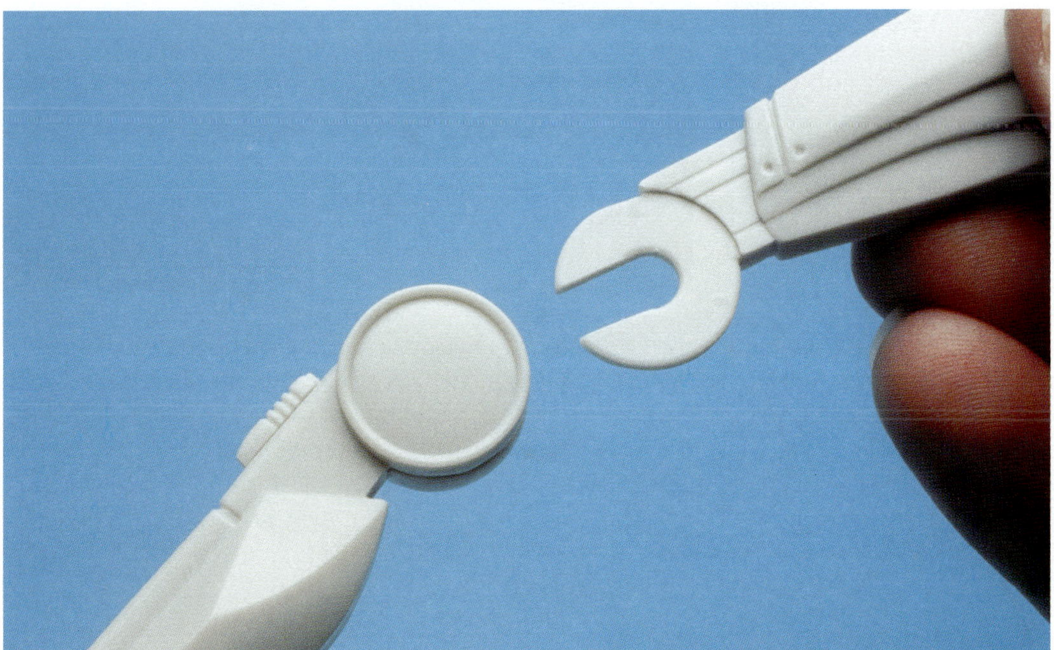

10 To make the seams easier to fill, I assembled the upper and lower legs separately. Afterward, I cut a slot in the upper leg so I could slide the parts together and adjust their fit later.

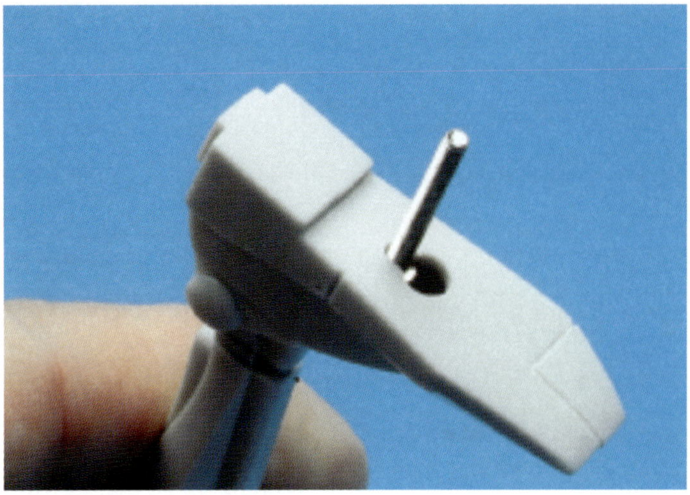

11 I doubted that the tall, skinny model would stand on its own, so I had to make a display base and find a way to attach the model to it securely. I had some small nuts and bolts that would work nicely, so I filed a notch into each of the model's feet as shown. The notch is shaped to hold the model to the base securely when the nut is tightened; the bolt can be removed altogether when the nut is loosened.

12 Here's a display base trick I learned from my friend Paul Boyer, *FineScale Modeler* magazine's senior editor. I found a plaque from a model contest I had entered a few years ago. The metal faceplate had a scratch in it from where I had dropped it, so I hadn't been displaying it. A faceplate attached to the wooden base with double-sided adhesive tape like this one can usually be pried loose with a straight-bit screwdriver. A little masking tape on the end of the screwdriver will keep it from scratching the wooden base.

13 Once you've lifted a corner of the faceplate, grab it with a pair of pliers and slowly peel it off. If the adhesive is stubborn, heating it with a hair dryer will help. Usually your efforts will reveal a pristine wooden base like this one that you can recycle back into your model collection. If you don't have any unwanted plaques around, bases like this one are available inexpensively at trophy shops and craft stores.

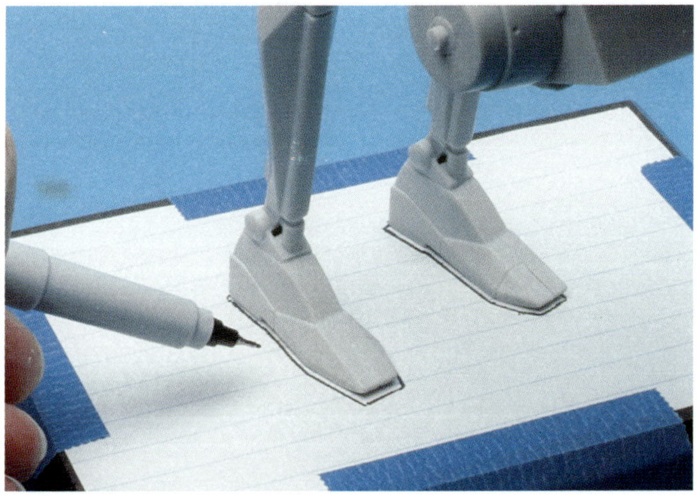

14 I needed to drill a couple of holes in the base for the mounting bolts. I taped a piece of notebook paper to the base, then positioned the model and traced around its feet with a marker. Next, I marked the location of the bolt holes on the tracing and drilled the holes using a motor tool. Bases like this one are usually made from processed wood; it's easy to drill through smoothly.

15 Now I could slip the model into place on the base and tighten it down from underneath.

16 Again using the motor tool, I countersunk the bolt holes under the base to make the nuts easier to tighten.

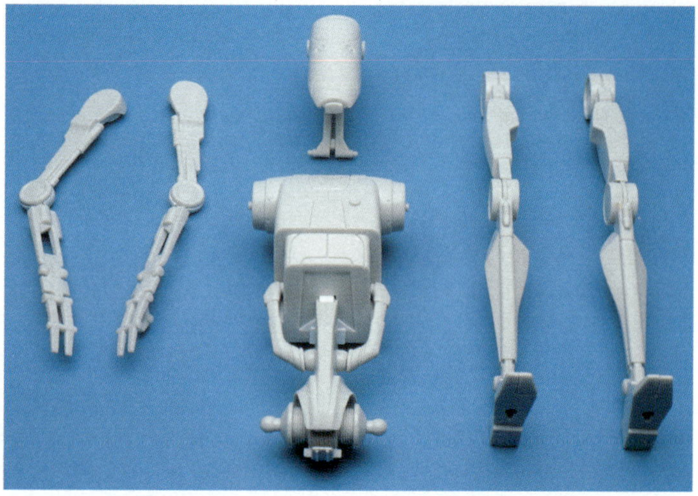

17 Let's get back to the model. The model's arms are molded in a wide-open pose so they can be attached to the STAP's hand controllers. The open arms look a little awkward when the droid is in a standing pose, however, so I cut partway through the elbow joints with a razor saw and repositioned the arms using super glue. I made each arm slightly different to give the model a more natural look.

18 Here are the model's subassemblies, ready for paint.

19 Visit the paint section of almost any hobby shop and you'll find a huge variety of paint colors, including several tans. I had three colors I thought might match straight from the bottle, but it's always hard to judge paint color in the bottle. I found a scrap of sheet plastic and applied a few drops of each color to it, labeling each one. When the paint was dry, I decided that Testor British Gulf Armor Light Stone (no. 2136) was the closest out-of-the-bottle match to the color of the film droids.

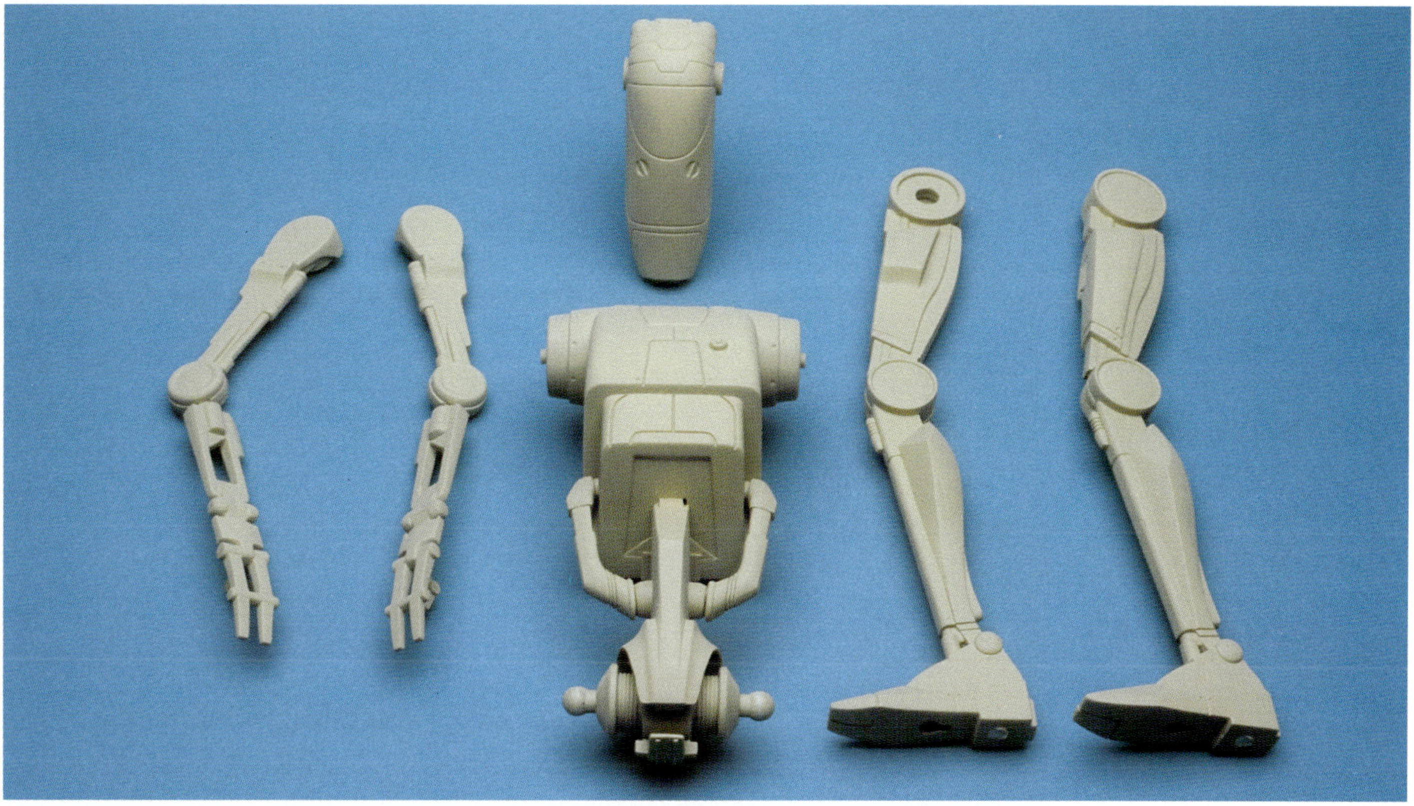

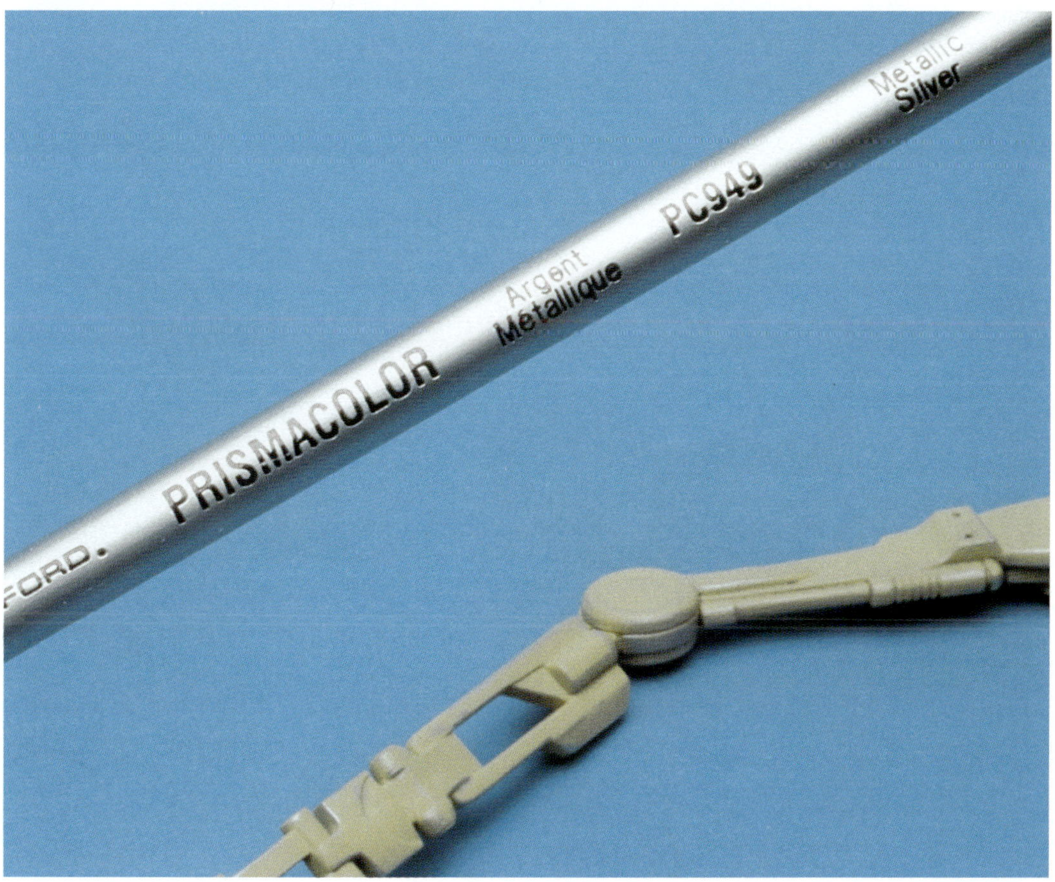

20 I airbrushed each of the subassemblies and allowed them to dry for several days before proceeding.

21 The Battle Droid should have a slightly worn finish where the metal bodywork shows through the tan paint. My friend Mike Fulmer showed me a great way to simulate this kind of wear using a silver pencil from an art supply store.

22 Sharpen the pencil, and rub the edge of the lead along the edge of the parts where you want to simulate wear. It's a simple technique that produces a realistic effect.

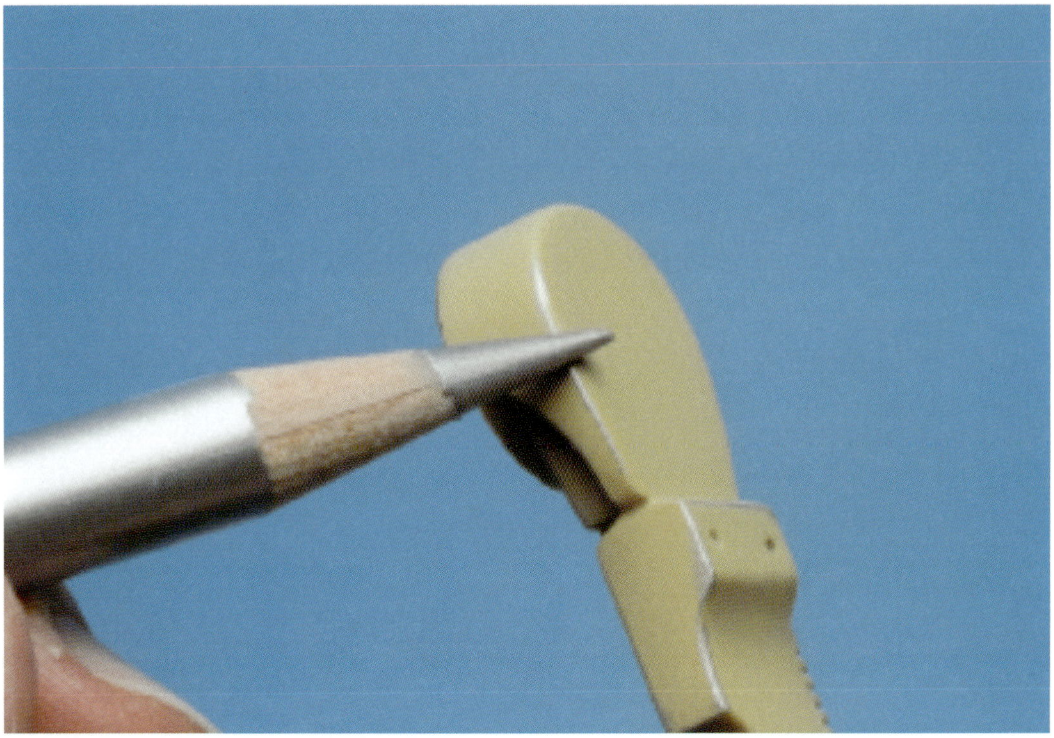

23 The model had a few larger areas where I wanted to simulate more wear. I drybrushed them with Floquil Old Silver paint (no. 110100) using a flat brush. To drybrush an area, dip the brush into the paint, then wipe almost all of it off on a sheet of paper or cardboard. When there's just a little paint left in the bristles, rub the brush along the edges where you want to simulate wear. It's best to use as little paint as possible, and let the effect build up with repeated applications.

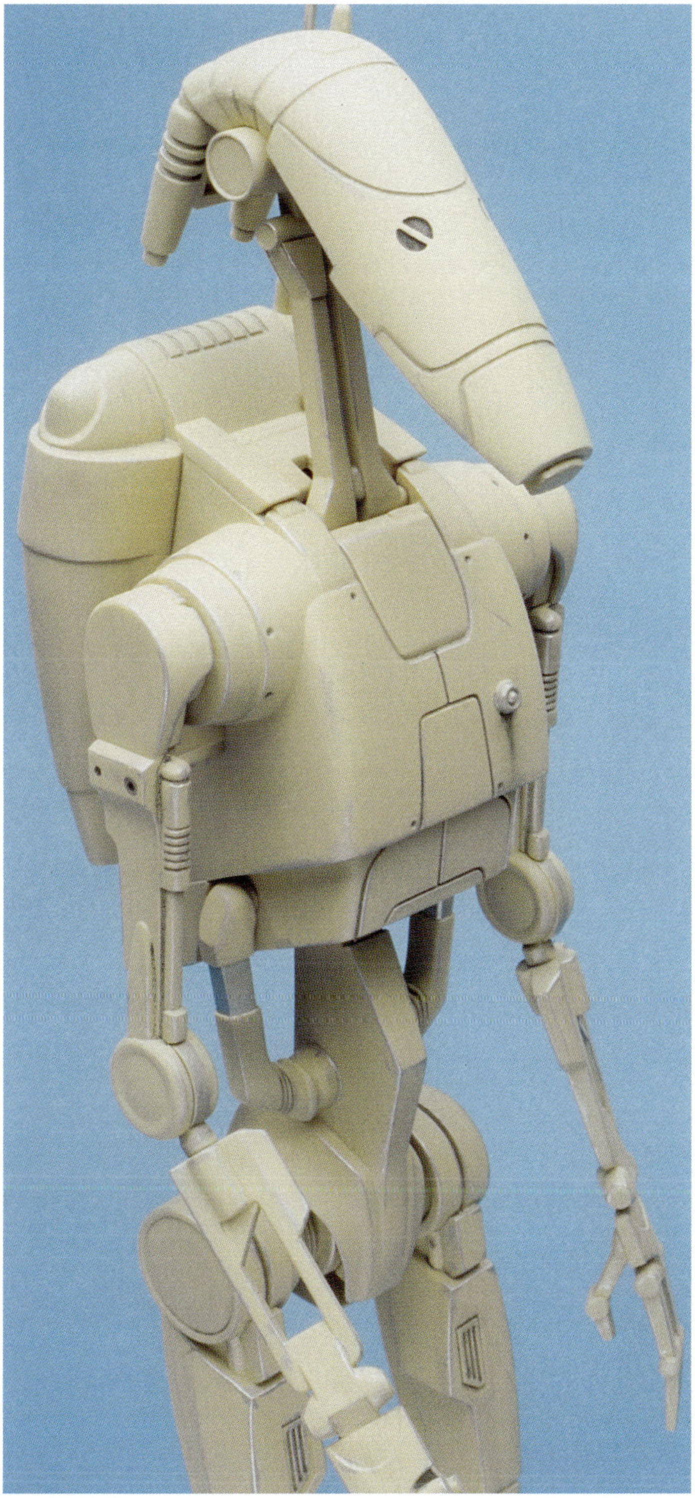

24 I used a soft lead pencil (also from the art supply store) to darken the model's panel lines. Afterward, I glued together all the subassemblies using gap-filling super glue. If you have a well-stocked spare-parts box and can do a little scratchbuilding, you could probably put together a pair or binoculars or a blaster rifle for the Battle Droid to hold without too much trouble.

25 Here's the finished Battle Droid installed on its display base. The striking model is more than a foot tall, and building it offers a great change of pace when you need a break from building spacecraft.

SEVEN

Millennium Falcon

In George Lucas' breakthrough film, *American Graffiti*, John Milner drives "the fastest car in the valley," a chopped-and-lowered 1932 Ford coupe. The bright-yellow car became an icon, and it was clear that George Lucas understood hot rods.

His understanding was even clearer when he included a hot rod—Han Solo's Millennium Falcon—in *Star Wars*. Sure, it's a science-fiction film, and the Falcon doesn't look anything like a deuce coupe, but it can be counted as a hot rod. Outfitted with a full complement of go-faster parts, the Falcon's speed and maneuverability make it one of the most important ships in the fleet—even if it won't win any beauty contests.

MPC's Millennium Falcon kit appeared soon after the release of *Star Wars*. Early versions even included a lighting system. A flashlight bulb illuminated the model's engine panels, cockpit, and gun turrets. The effect was pretty toylike, and the lights were eventually removed from the kit.

Although it's been released in several versions, the Falcon has always been a mixed bag—some details are spot-on, while others are way off. The model's proportions have always been one of its biggest problems. A resourceful modeler, Jack Smith, recognized that the kit's side panels—the mechanical details sandwiched between the ship's curved upper and lower sections—are too tall. Jack scratchbuilt thinner panels and offers cast-resin copies through his company, Falcon Kits. Jack also offers parts to upgrade the kit's cockpit and landing gear.

With an opportunity to correct the kit's proportional and detail problems, I also decided to light the model using LightSheet. LightSheet is a flexible plastic film that glows like neon and doesn't generate any heat, so it's perfect for lighting the Falcon's cockpit and engine panel.

Let's dig in and build an accurate replica of the galaxy's fastest Corellian cruiser.

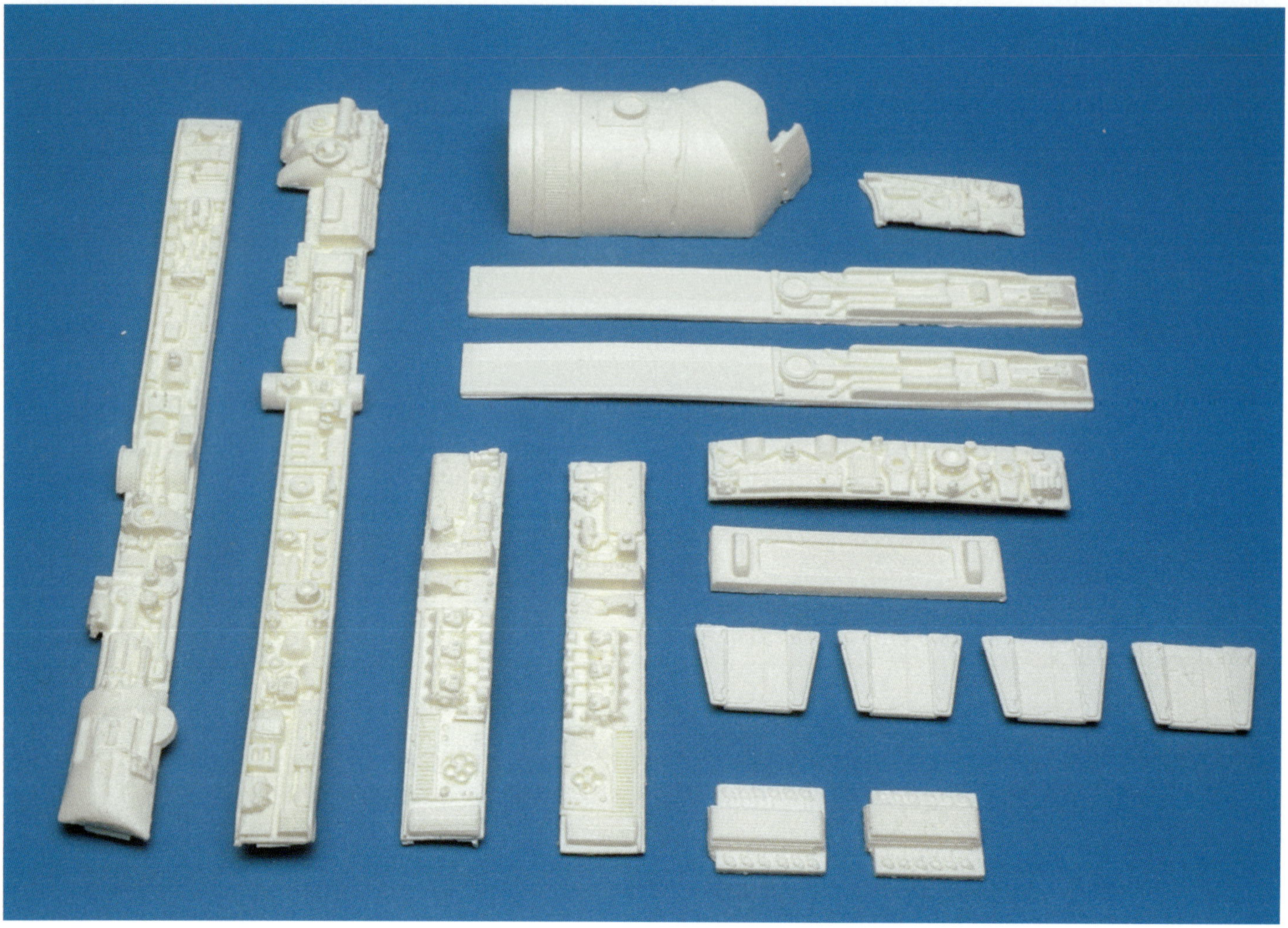

1 Here's a look at Falcon Kits' cast-resin side panels. The parts are cleanly cast and replace the kit's too-tall plastic parts without too much trouble.

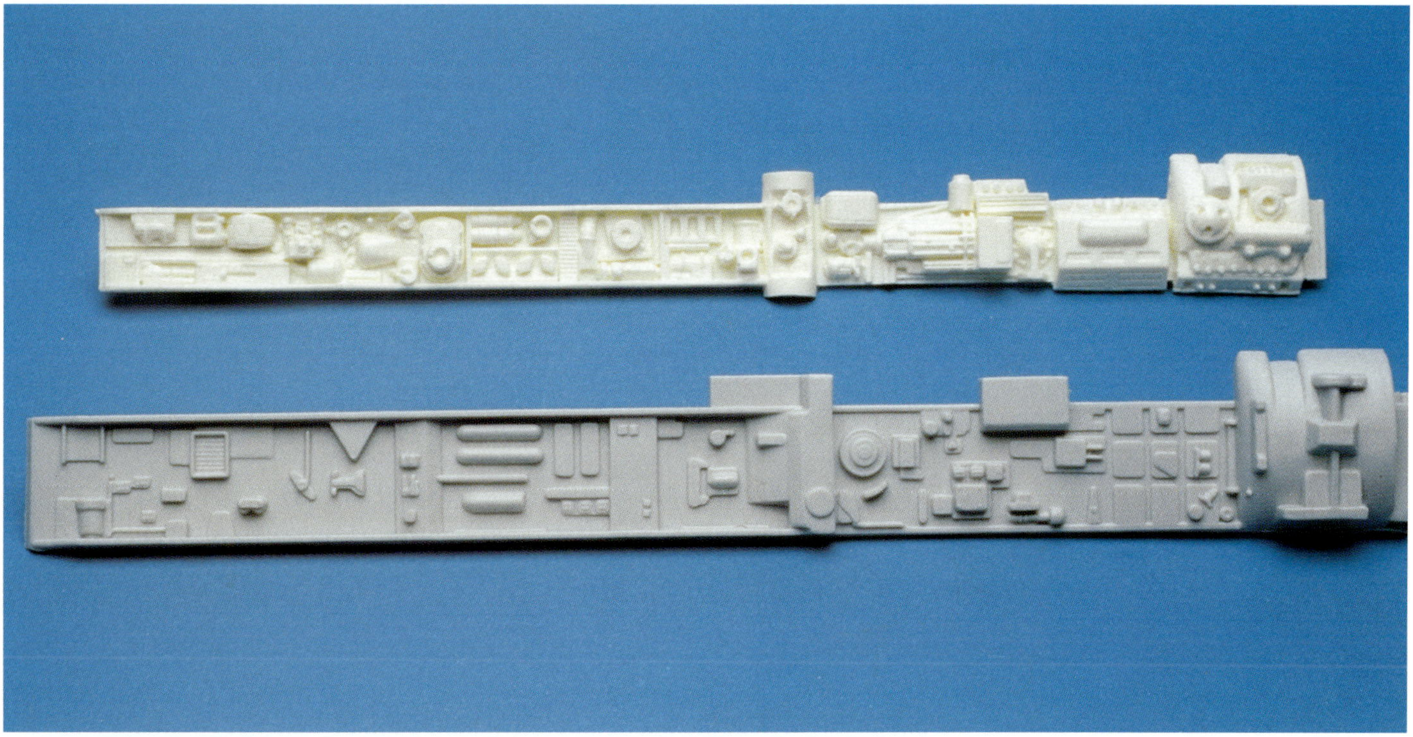

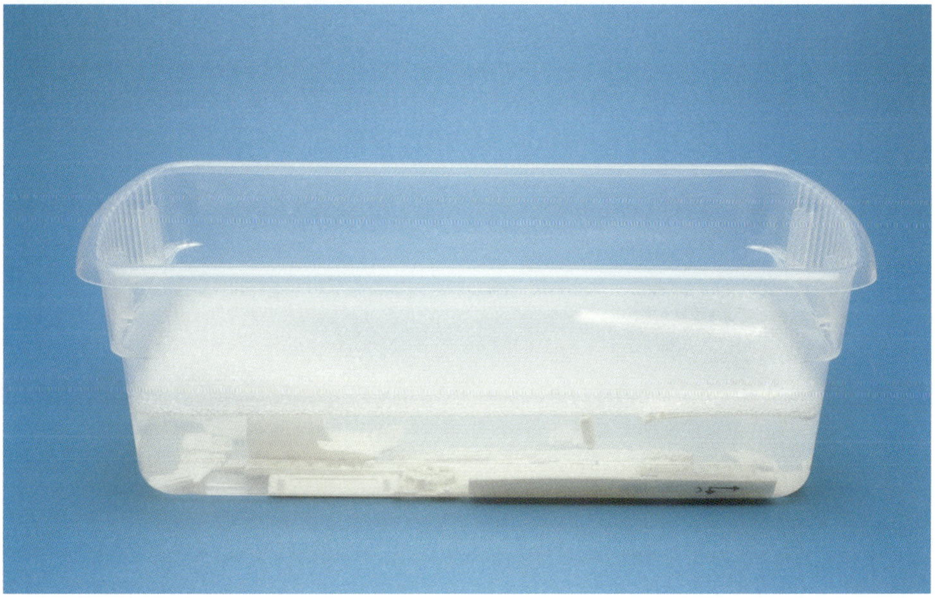

2 The Falcon Kits cast-resin side panels correct the kit's too-tall plastic panels and are considerably more detailed.

3 Cast-resin parts are often coated with a strong mold-release agent that can wreak havoc on adhesives and paint. Before you begin assembly, wash all the resin parts in soap and water. Let them soak for a half-hour or so, then give them a quick scrub with an old toothbrush. Rinse the parts thoroughly in a second container filled with warm water. Don't use the sink—little parts can accidentally disappear down the drain. (I'm speaking from experience here.) Allow the parts to dry thoroughly before proceeding.

4 Let's start on the lower hull. It's plagued by a variety of crude operating features that need a lot of work to correct. Falcon Kits provides new forward landing-gear bays that are applied to the outer surface of the hull. The kit's hull has large openings where the ship's mechanical details show through, but they're where the bays need to be installed. Inserts make the holes in the hull smaller so the bays can be added. Glue the inserts in place with gap-filling super glue or epoxy—regular model glue won't work with cast resin.

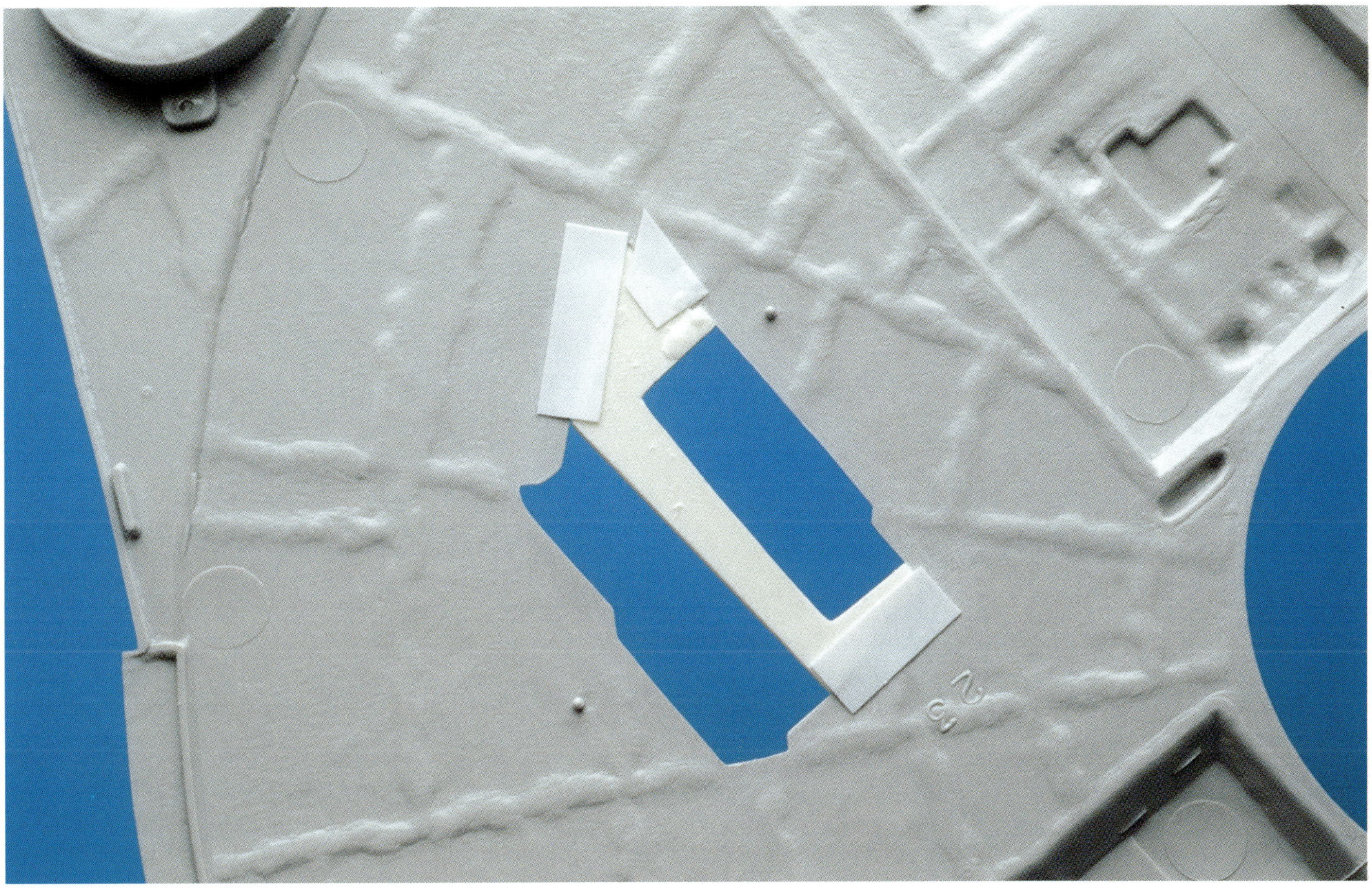

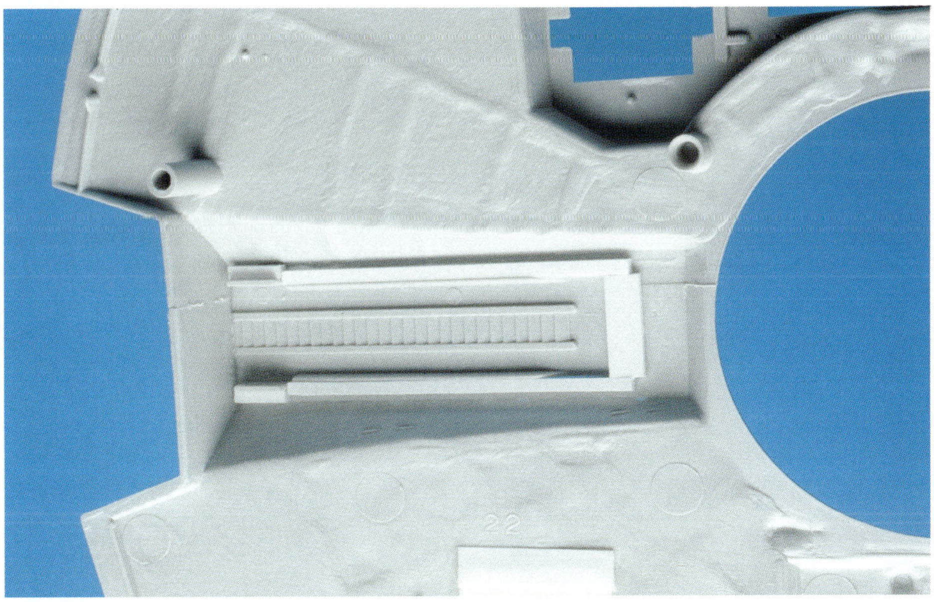

5 I reinforced the inserts inside the hull by gluing strips of .010″ styrene sheet over the joint between the parts. Despite the modifications, the kit's mechanical detail bays (parts 22 and 23) are installed as usual from inside the hull.

6 The model features a working boarding ramp. I cut off its hinge pins with a razor saw and glued it in place using gap-filling super glue. Again, I glued styrene strip over the joint between the parts.

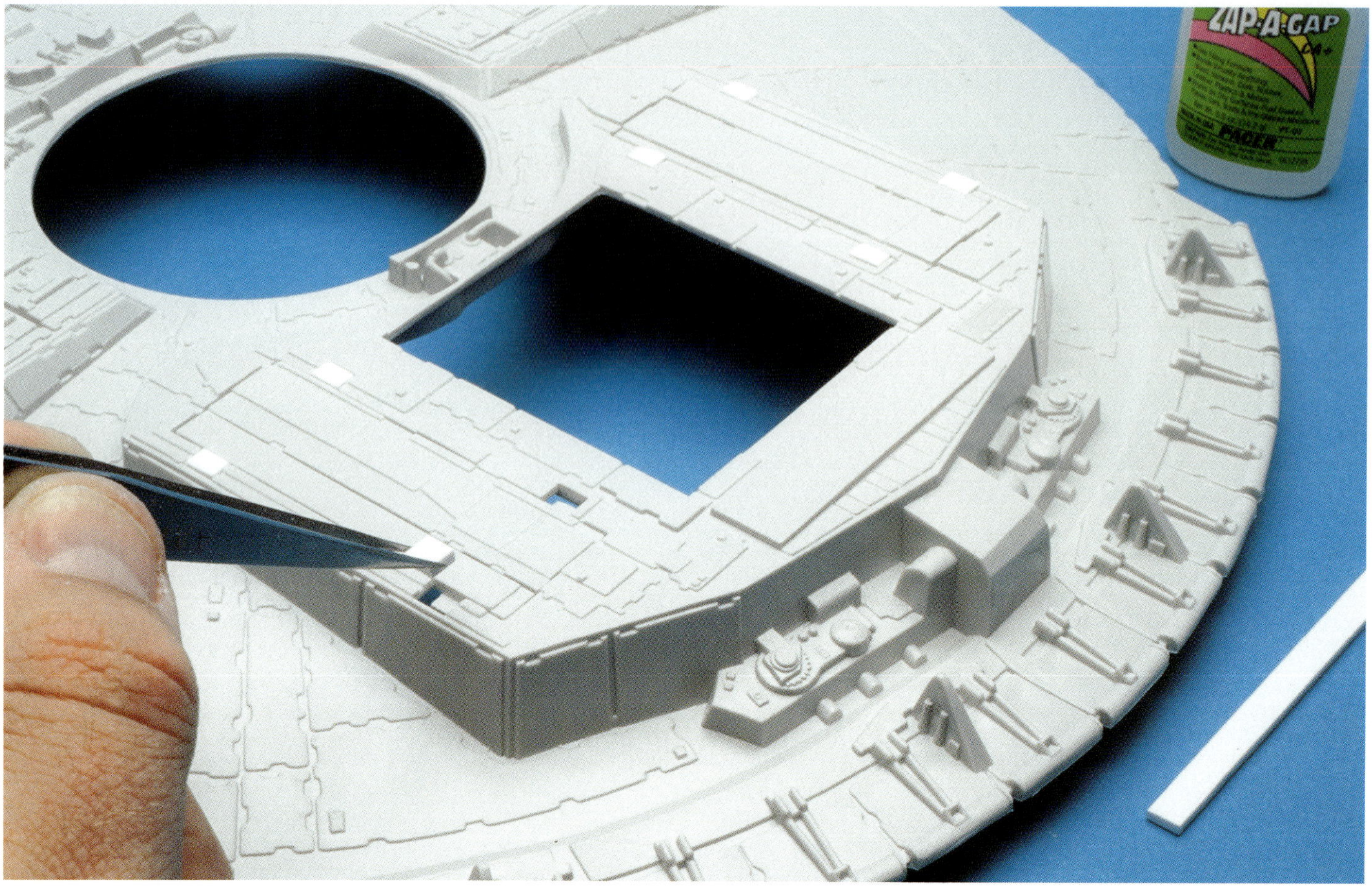

7 The kit has a crude set of working landing gear, too, with hinged gear doors. Since I was going to model the Falcon in flight (with its gear retracted) I cut off the hinges and glued the doors in place. Rectangles of sheet styrene fill the hinge openings.

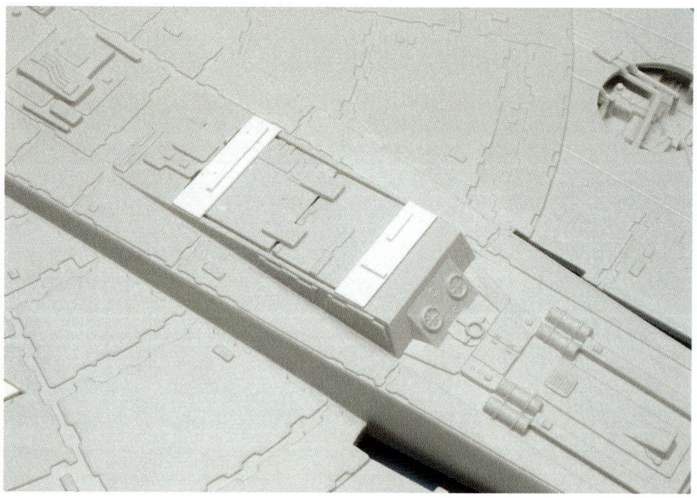

8 The doors on the kit's single forward landing-gear bay fit terribly. I added sheet-styrene panels on the outside of the hull to help disguise the huge gaps between the parts.

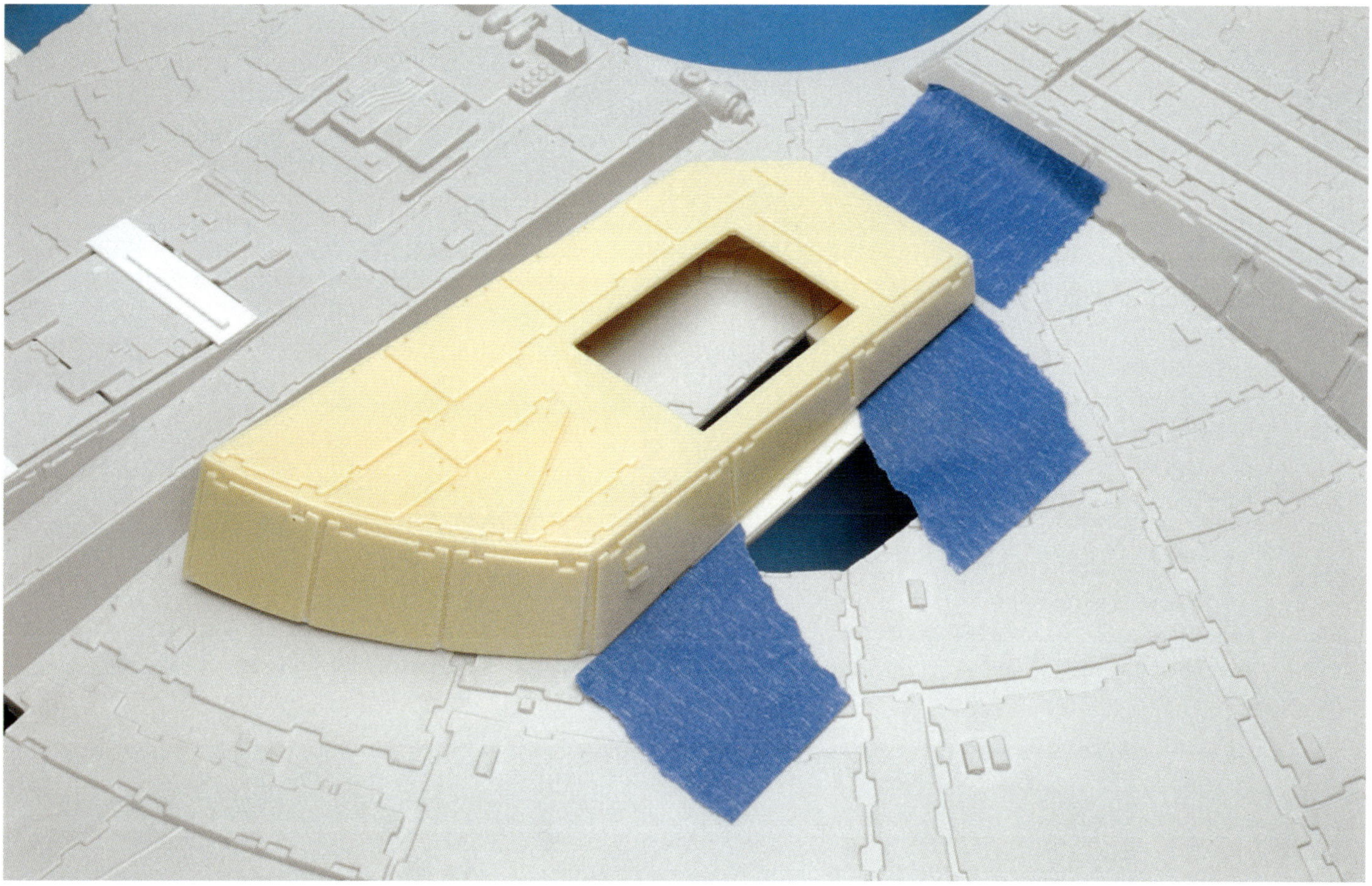

9 Next step Is to add the external landing-gear bays. Falcon Kits provides a full-size template to help locate them correctly on the hull. I marked the locations with pieces of masking tape and glued the parts in place. Afterwards, I added the landing-gear doors.

10 Here are the new landing-gear bays in place on the model.

11 Although it's no longer mentioned in the instructions, the kit includes a large compartment that dates to the original version of the kit. It held the two D batteries that powered the flashlight-bulb lighting system.

12 The battery compartment is accessible through an access panel in the bottom of the hull.

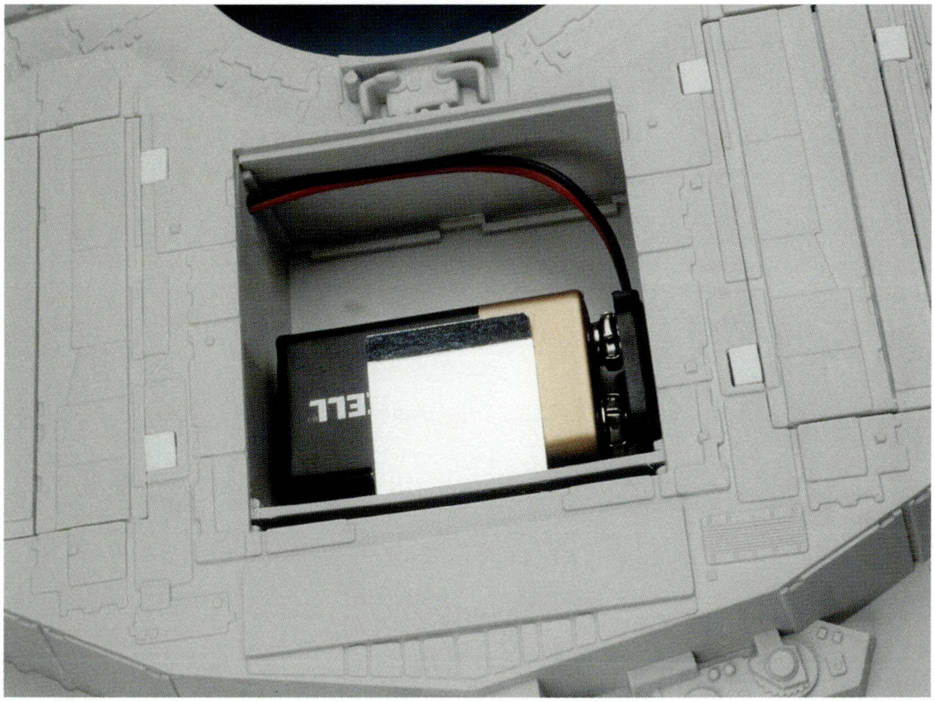

13 The battery compartment is large enough to hold the 9-volt battery that powers the LightSheet system. I found the battery holder at Radio Shack; it's bolted to the compartment wall. Another one holds the two AA batteries that power the headlights.

14 Correcting all the lower hull's problems takes plenty of time, but makes for a much more detailed (and much less toylike) model.

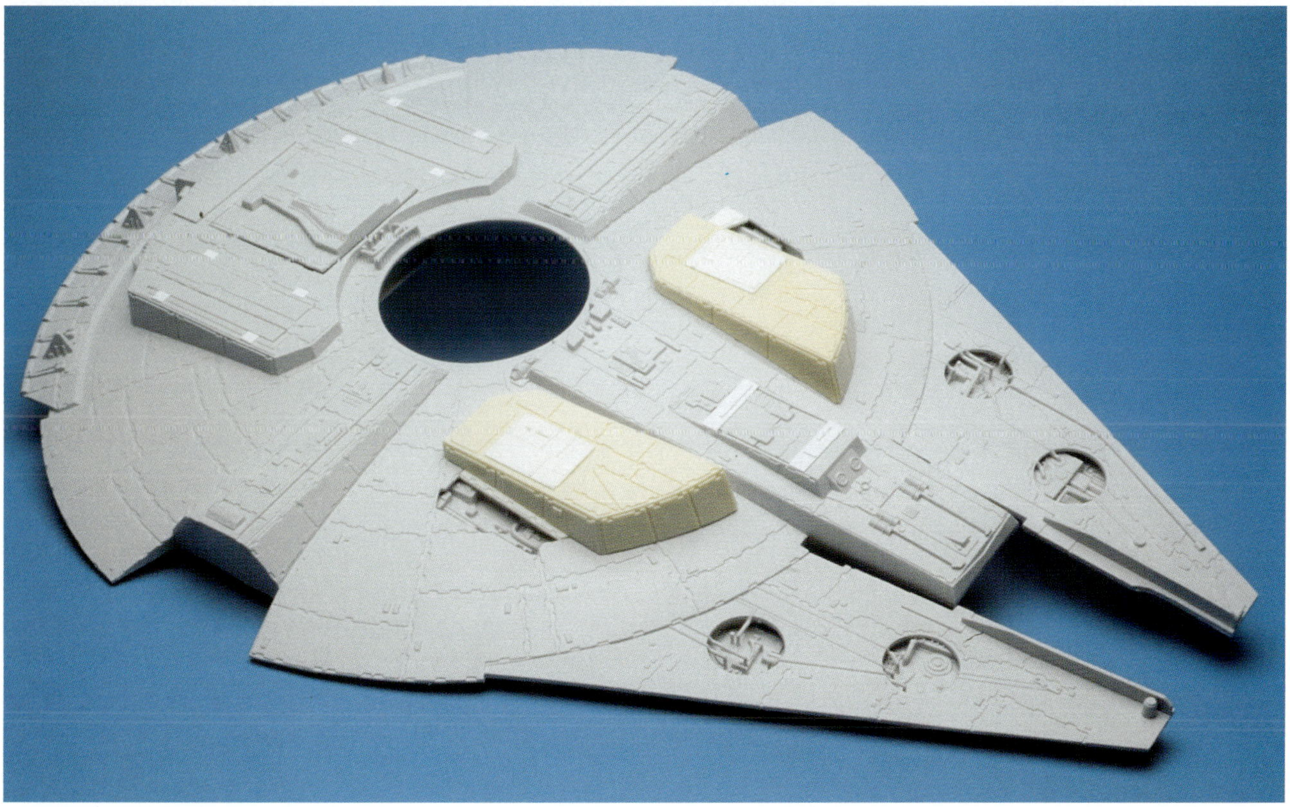

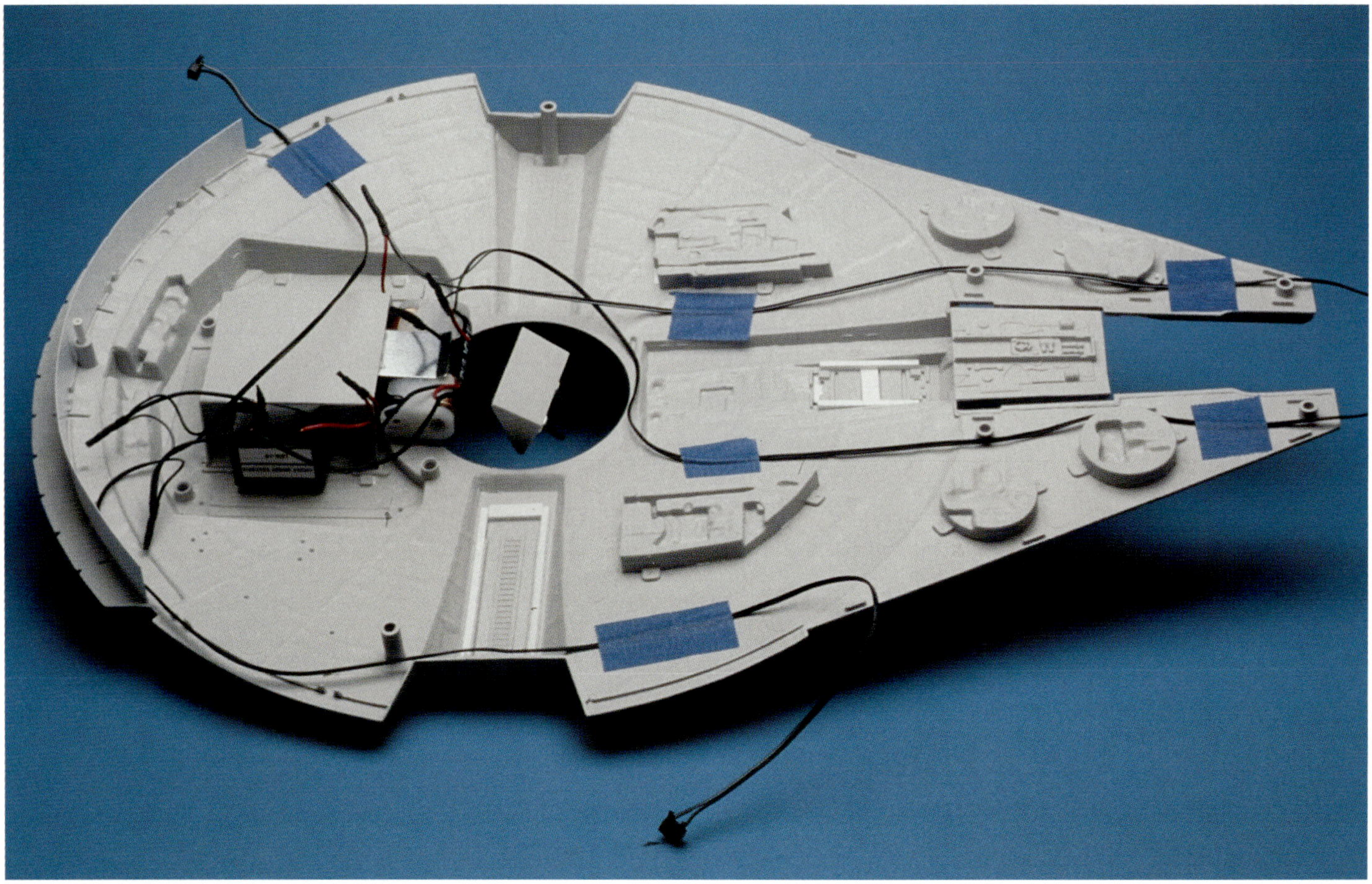

15 Now is a good time to install the lighting-system wiring. There's plenty of room inside the model for all of the components. I used masking tape to hold the wires in place as I mapped out all the circuits. When I was sure everything worked properly, I soldered the connections together. I added a 3/4"-tall sheet-styrene panel at the rear of the model to hold the strip of blue LightSheet.

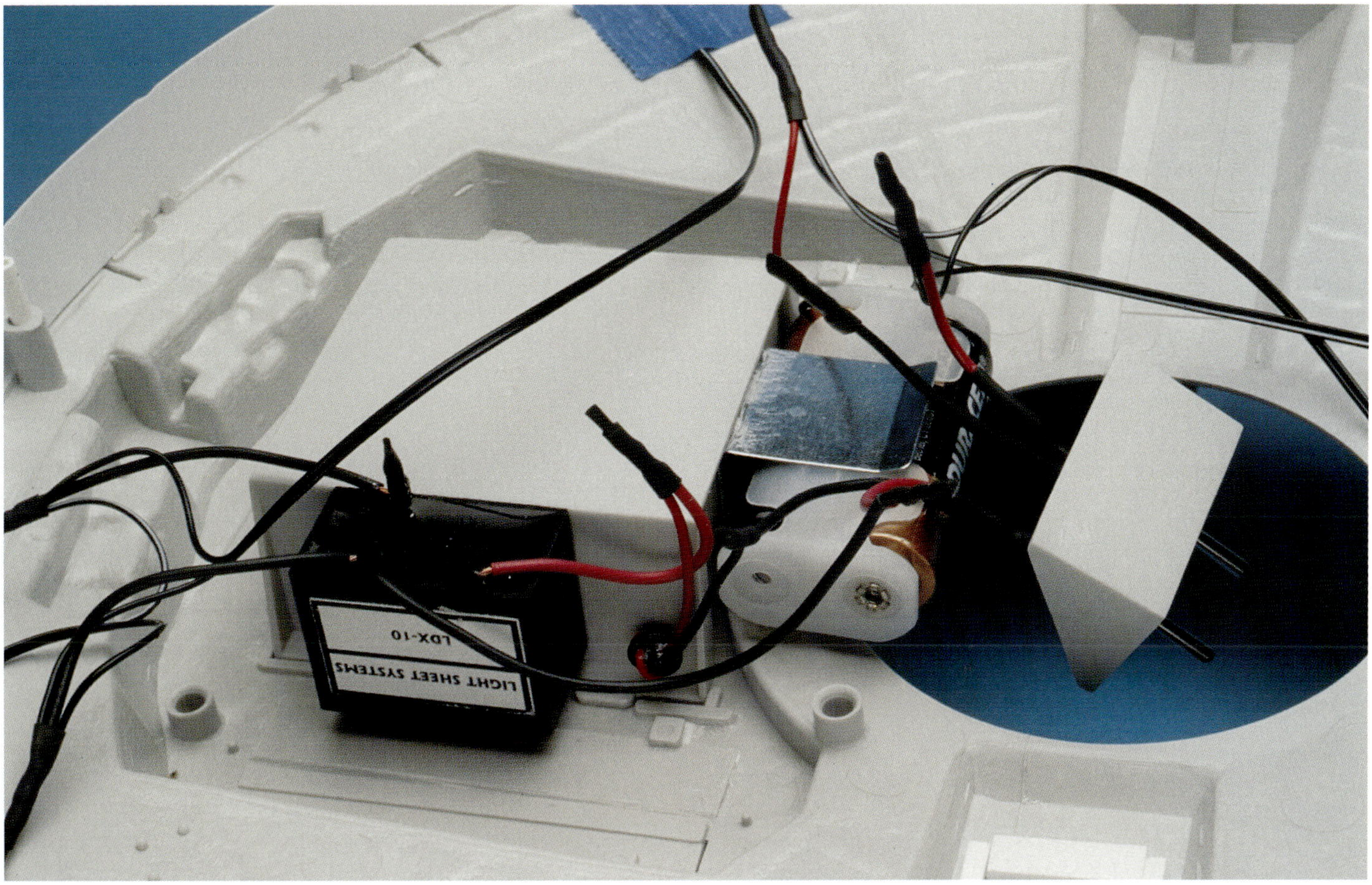

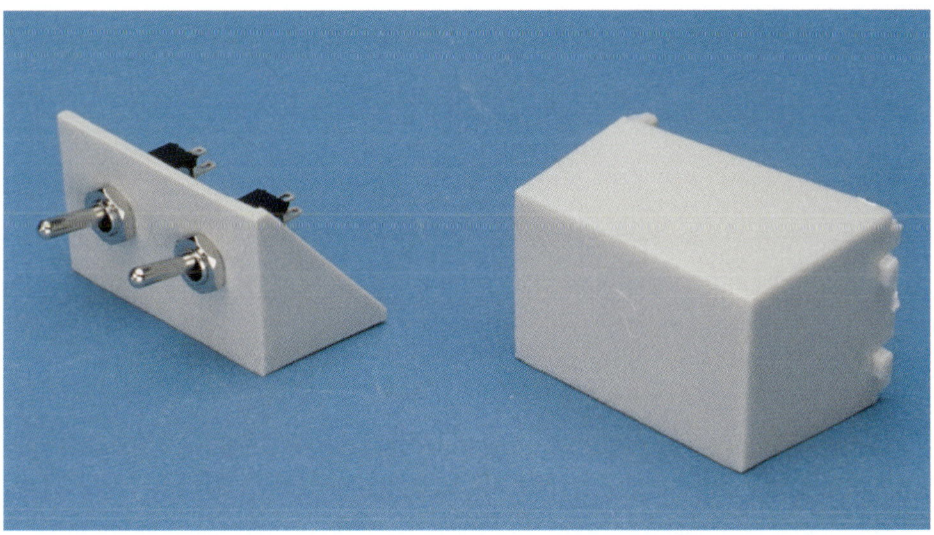

16 To prevent short circuits, I insulated all the soldered connections with heat-shrink tubing. The battery compartment is the backbone of the lighting system. The LightSheet converter is glued in place on one side, and the AA battery holder is bolted to the front. When I need to replace the batteries, I can access them by removing the round lower gun turret.

17 Two toggle switches control the lights—one for the headlights, and one for the cockpit and engine lights. I mounted the switches in a control panel I made from the kit's discarded forward landing-gear bay (part 72).

18 Let's move to the upper hull and install the resin side panels. Falcon Kits provides a map that shows where all the parts should be installed. Some of the curved panels may have flattened slightly. Don't force them—heat them for a few minutes in warm tap water and they'll be easier to bend. Adding a fillet of gap-filling super glue along the edge of each part strengthens the model considerably.

19 Working your way around the edge of the hull, install all of the side panels. Adjust their fit as necessary as you go. Falcon Kits doesn't supply parts to replace the ship's two round loading ports, but the kit parts will fit with just a little modification.

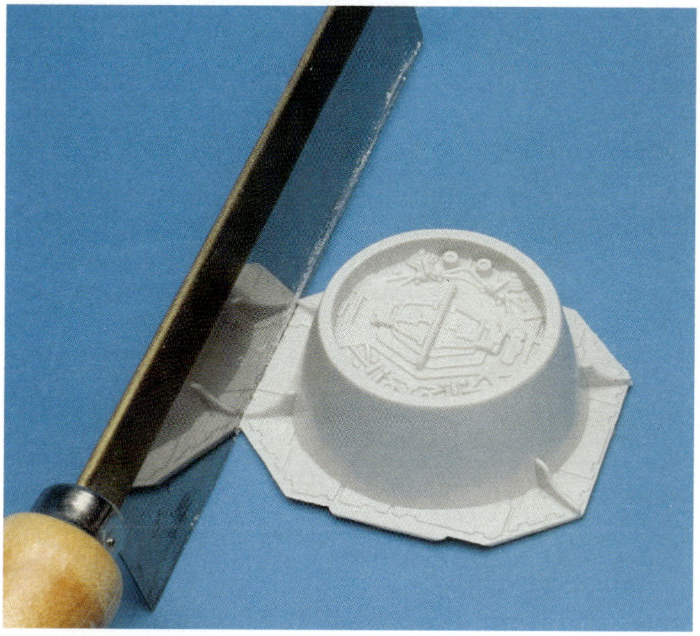

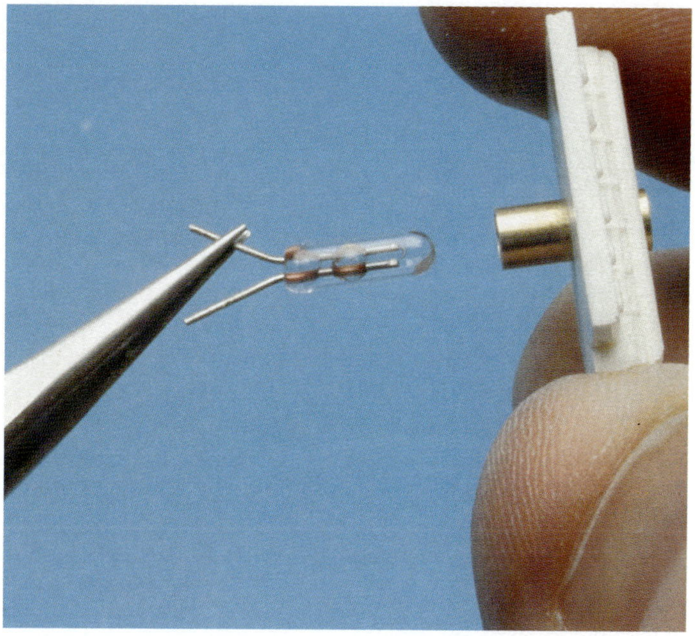

20 Trim a little plastic from the top and bottom of the docking ports, and they'll fit with the new side panels.

21 I wanted to install krypton flashlight bulbs in the front of the model to simulate the headlights. I drilled holes through parts B and F, and glued in short sections of brass tubing to hold the tiny glass bulbs.

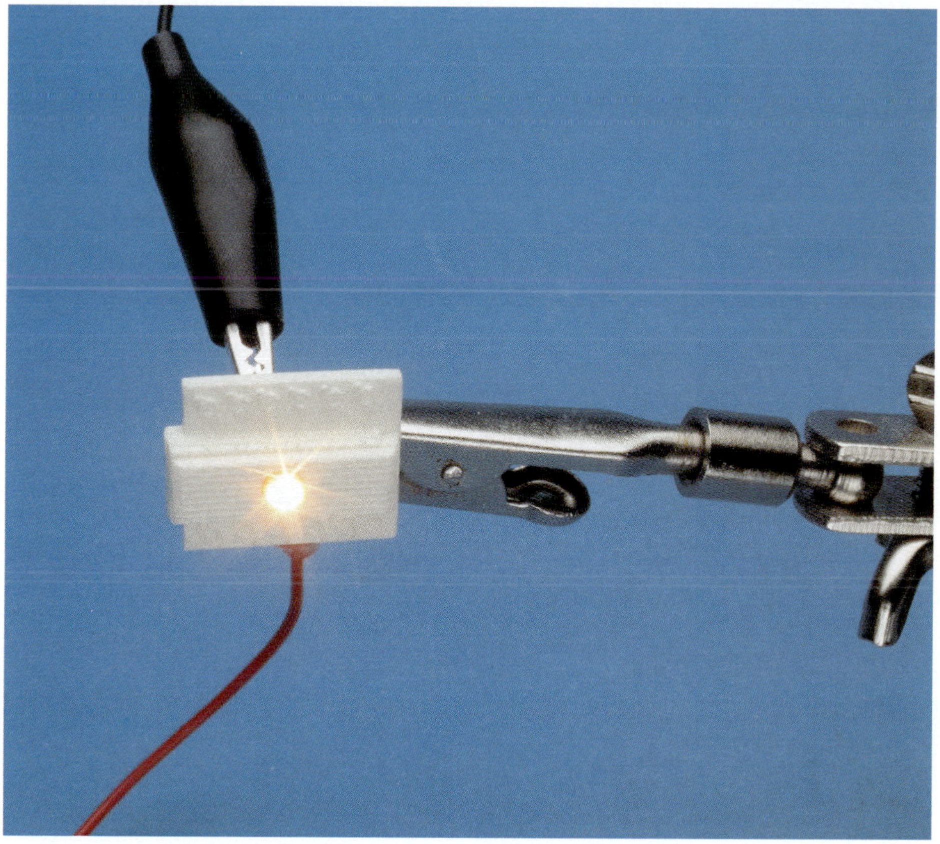

22 The brass-tubing bulb holders prevent light leaks through the sides of the model and dissipate some of the heat generated by the bulbs.

23 Before you join the upper and lower sections of the model, carefully glue the LightSheet engine panel to its sheet styrene backing and hook up its wiring. Test-fit the upper and lower hull halves, then make sure the engine still lights properly. Now is your last chance to check for short circuits. When you're satisfied that everything is working as it should, glue together the upper and lower halves, using gap-filling super glue.

24 When the halves were joined, I glued the switch panel to the inside edge of the upper gun-turret opening.

25 The only side panels I didn't install were the headlight panels, parts B and F. Leaving the holes open made it easier to glue together the slender forward sections of the hull—I was able to add reinforcing fillets of super glue inside the model through the openings.

26 After the hull was assembled, I soldered the krypton bulbs to their wires and installed the headlight panels.

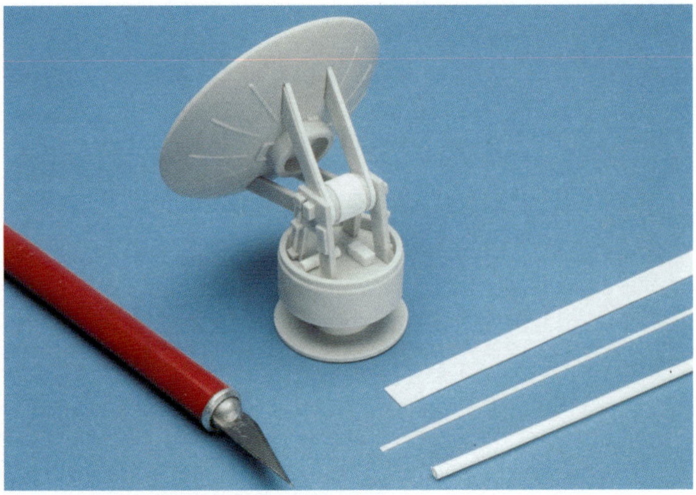

27 The upper sensor dish is poorly detailed straight from the box, so I added some styrene strip and tubing to dress things up a little.

28 Detailing the cockpit would make it one of the model's focal points, so I needed to do something to dress up the kit's spartan forward canopy (part 6). Comparing it to the box-art photo of the studio model shows that it's missing some vertical braces and could use some raised detail around the edges of the window panels.

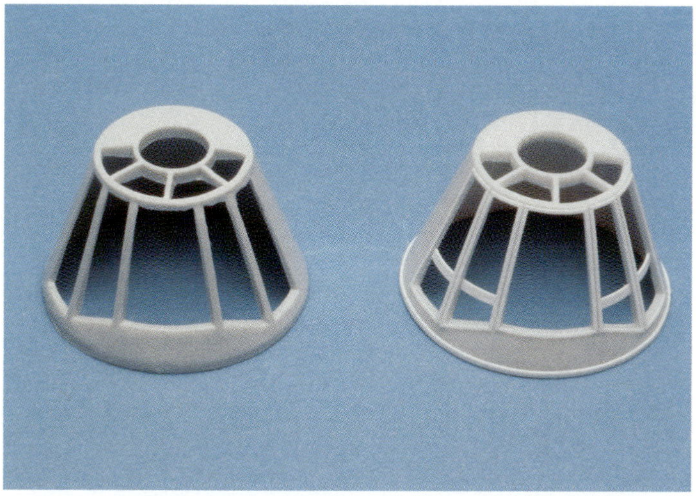

29 Using styrene strip, I added the missing vertical braces and window framing. Compare the original kit part (left) to the modified part.

30 When all of the subassemblies were ready, I gave the model an overall coat of Testor Model Master Light Gray (no. 1732) using an airbrush. The Falcon is a big model—it took almost three ½-ounce bottles of paint to cover all of it. Note that I haven't installed any of the cockpit parts yet. The lower half of the cockpit tube is a separate part that can be installed later on along with the rest of the components.

31 If its fuselage is any indication, the Falcon has been repaired with parts from a couple of different ships—some of its mismatched exterior panels are dark red, and some are dark gray. I airbrushed the dark gray panels using Testor Model Master AMC Gray (no. 2035), masking them along the molded-in panel lines. Afterward, I did the same for the red panels using Testor Model Master British Crimson (no. 2009).

32 To break up the rest of the gray paint scheme, I darkened some of the Light Gray paint and used it to darken the panel lines and some of the individual hull plates. I also used it to darken the finish inside the open mechanical areas of the hull.

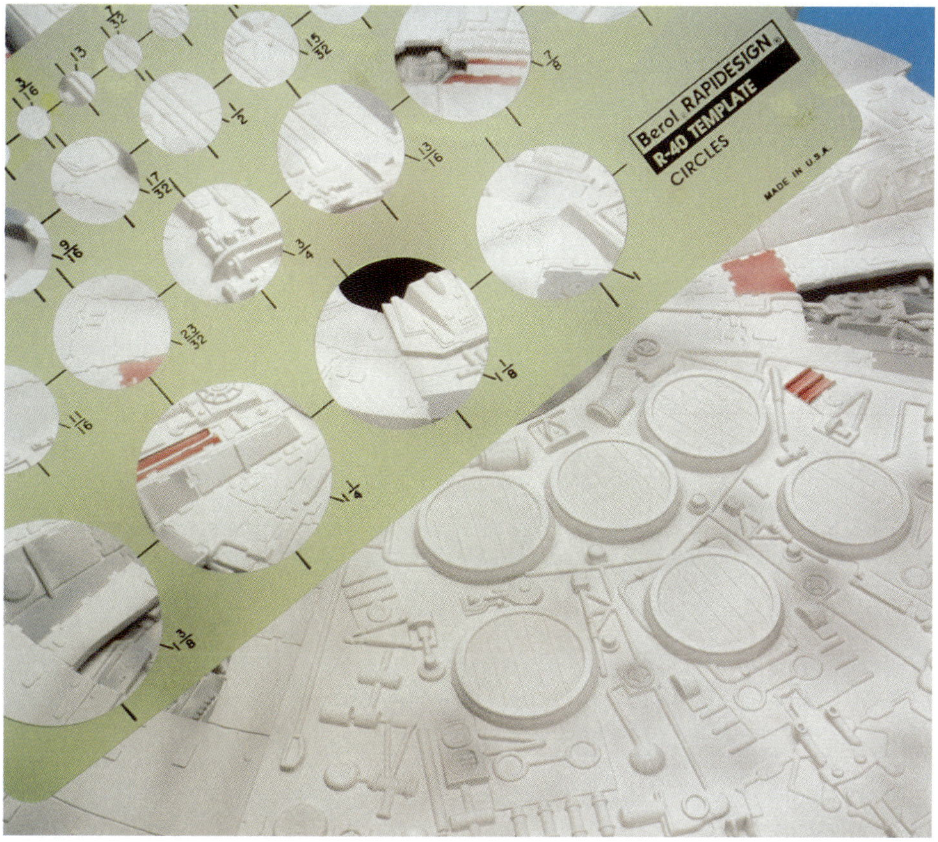

33 The Falcon has six round engine grilles on its rear deck. I cut a circular mask from frisket film using a circle template and a sharp hobby knife, then airbrushed the grilles flat black.

34 Clearly the Falcon is burning a little oil! After I painted each of the grilles flat black, I used the airbrush to add the heavy black streaks that trail off the back of the ship.

35 While the weathering is drying, let's do some work on the cockpit. Here are the parts in the Falcon Kits cockpit set.

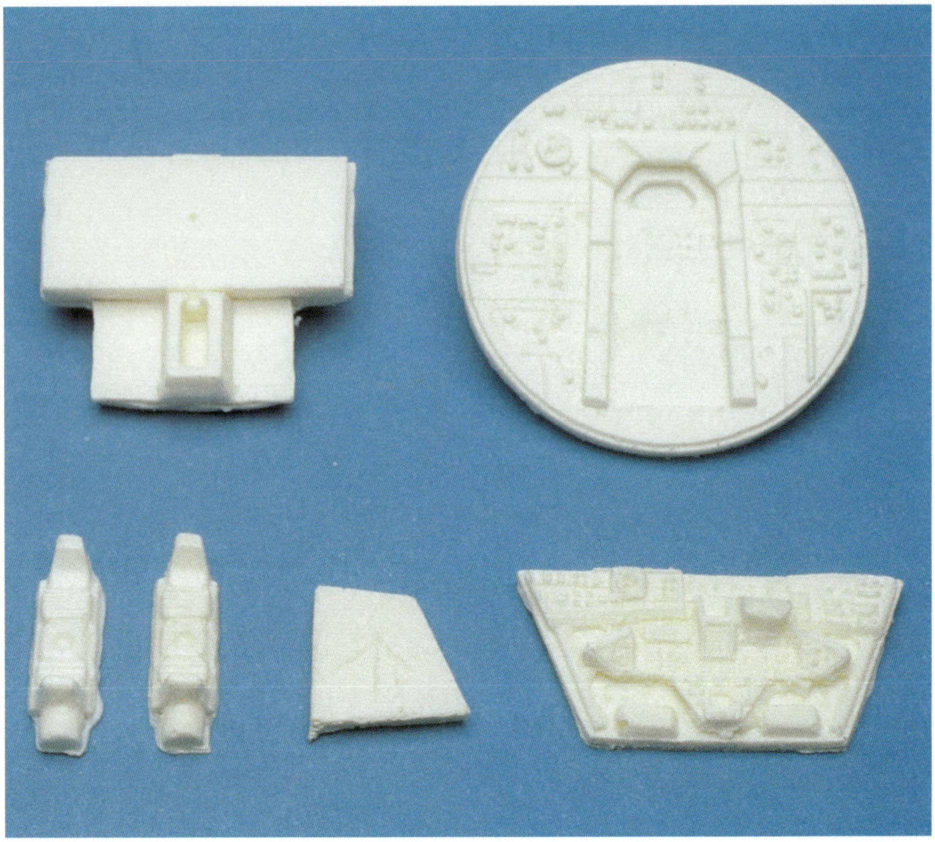

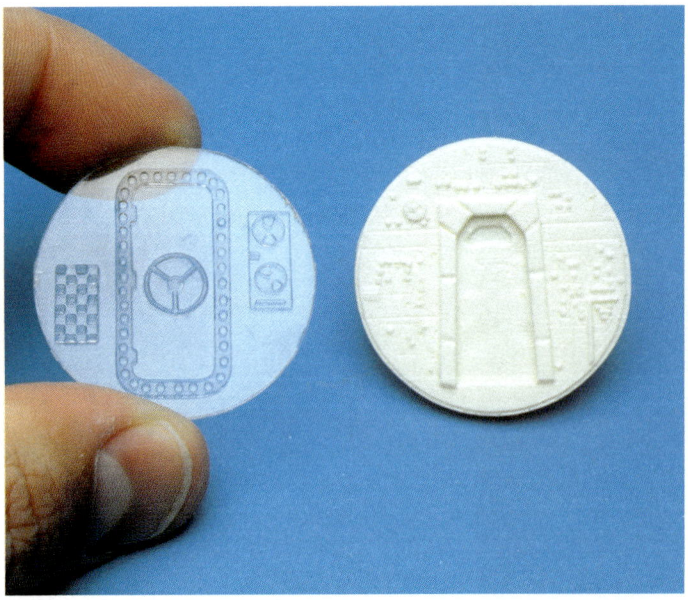

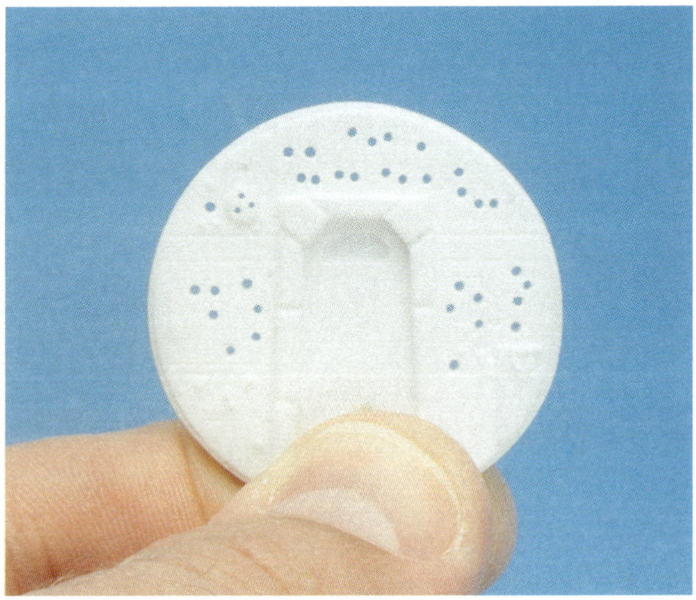

36 The rear bulkhead in the Falcon Kits set is a vast improvement over the "submarine hatch" that comes with the kit!

37 The rear bulkhead is covered with lights, and I wanted to include them on my model. I chucked a tiny bit in my drill press and drilled a couple dozen holes through the cast-resin part, using its molded-in lights as a guide. Next, I cut out the horizontal light panels using a razor saw.

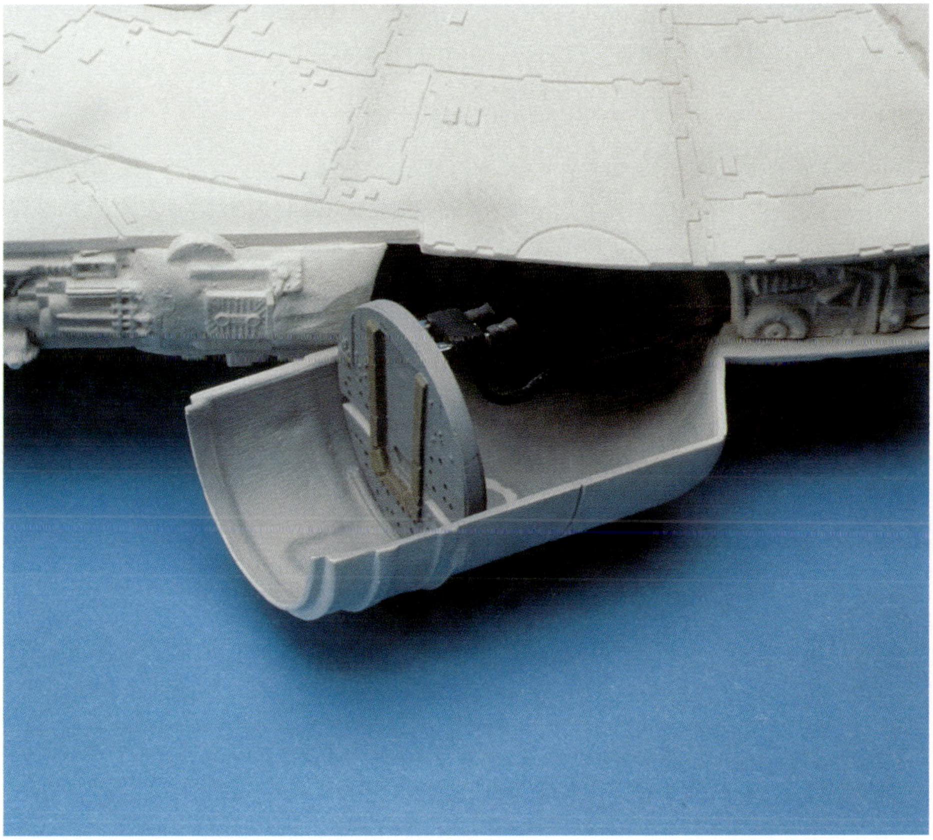

38 When I planned the lighting system wiring, I added a second, smaller LightSheet panel to illuminate the cockpit bulkhead from behind. White LightSheet is bright pink when not energized.

39 I placed the bulkhead on a white LightSheet panel and traced around it with a hobby knife. Glued to the back of the bulkhead, the white LightSheet shines through the holes and illuminates everything evenly.

40 I glued the finished cockpit in place on the upper cockpit tube, attaching it behind the molded-in locator rib. The connector tabs on the LightSheet would have interfered with the lower cockpit tube; I bent them back carefully using needle-nose pliers, plugged in the wiring connector, and tested the circuit.

41 When I was sure the rear bulkhead was installed correctly, I added the lower cockpit tube and the new interior. Three figures are included in the kit, but they're poorly molded, and only Chewbacca is worth adding to the model. (Maybe Han is in the back, working on the hyper-drive.)

42 Finally, it was time to add the forward canopy. The injection-molded "glass" included with the kit is far too thick and won't work with the Falcon Kits instrument panel. Using the template Falcon Kits provides, I cut new windows from clear sheet plastic with a sharp pair of scissors. I installed them inside the canopy using white glue. I let the glue dry overnight, added the instrument panel, and installed the canopy.

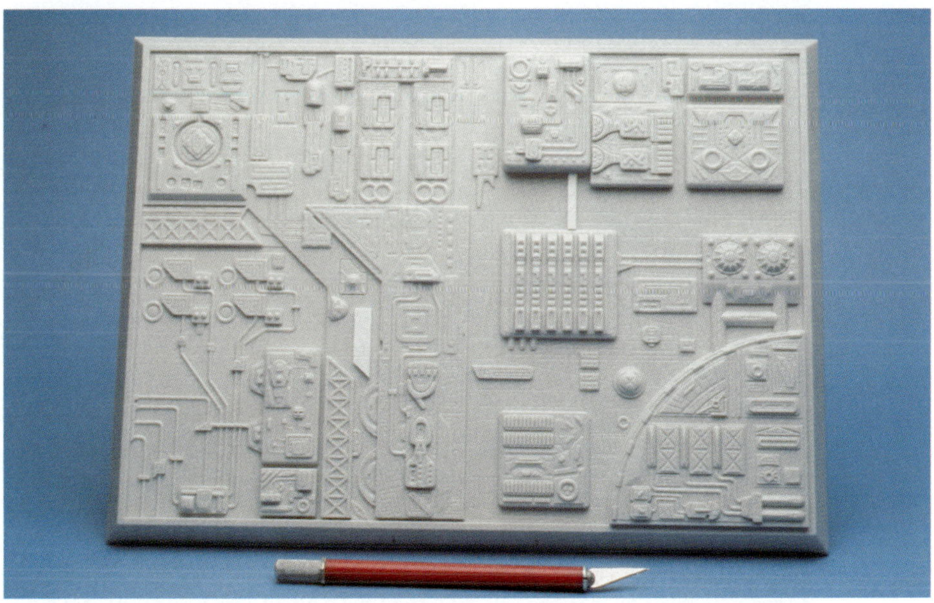

43 The Falcon was finished, but how would I support the "in-flight" model?

44 Looking at my spare kits, I realized that I still had the Death Star display base that came with the Imperial TIE fighters kit. At 8½" x 11½", it was large enough to fit under the model. I covered its molded-in stand slots with scrap bits of sheet styrene.

45 The model is supported on three pieces of 3/8"-diameter acrylic rod I bought at a plastics-supply store. I drilled through the base using a drill press, added the rods, and secured them underneath using epoxy. Two of the rods fit into the recessed mechanical detail bays on the bottom of the ship. The third rod supports the rear of the model near the battery-compartment door. I airbrushed the base Testor Model Master Gunship Gray (no. 1723).

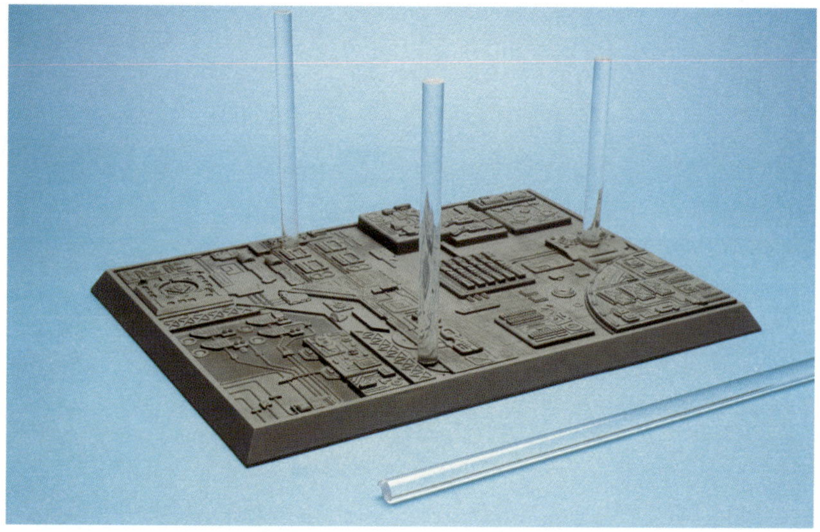

47 The bright blue LightSheet makes an incredibly realistic engine on the Falcon.

46 Seeing the bulkhead lights blazing makes all the extra cockpit work worthwhile.

Tools

Most of the chapters in this book jump right into building the model. That's probably OK if you've been building for a while and have a few projects under your belt. But if you're just starting out in the hobby or only build occasionally, some of the tools, materials, and building techniques may be a little unfamiliar. Here are a few workbench tips that I hope will add to the material presented in the chapters and help you build better models.

A simple single-action airbrush like this Badger 200 is one of the best tool investments you can make. With it, you'll be able to apply smooth, even finishes to your models, custom-mix colors, and take on more advanced weathering techniques.

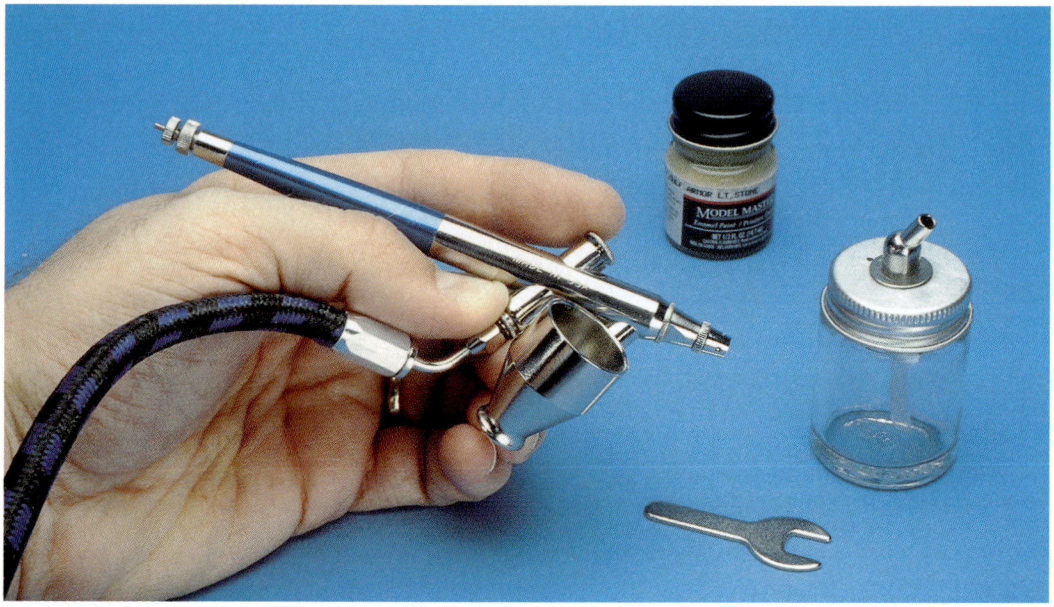

I painted and weathered my Millennium Falcon model using only a single-action airbrush. Despite its simplicity, it's a tool that's suited to a variety of painting tasks.

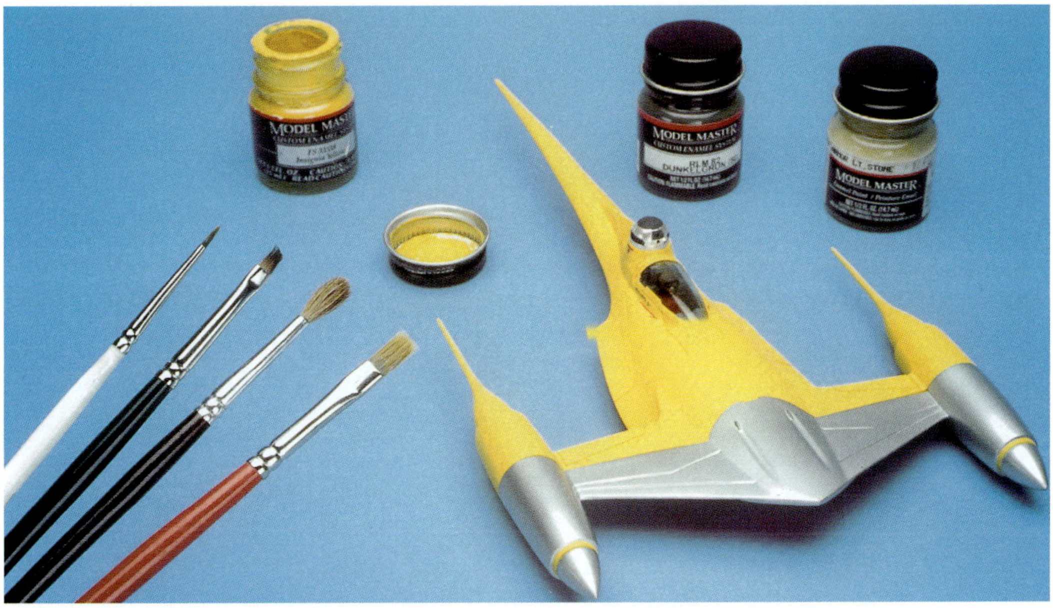

Keep a variety of paintbrushes handy for detail work. Clean them thoroughly after each use, and they'll last a long time.

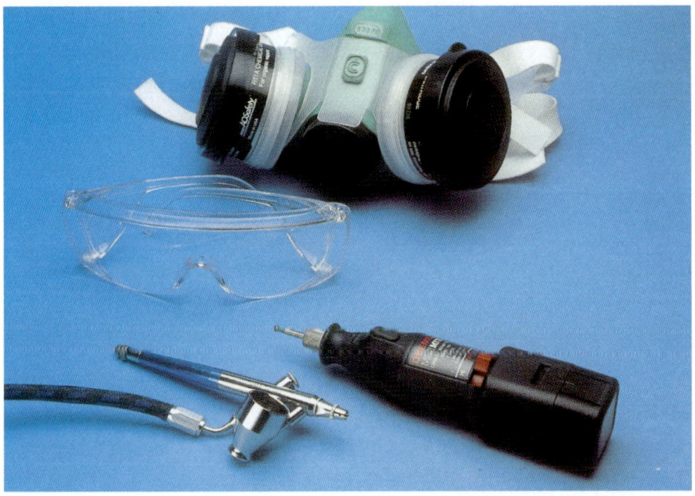

Safety should always be a priority when you model. If you use an airbrush, invest in a good two-stage filter mask like the one shown, and wear it every time you paint. They're available at most hardware and home-improvement stores. Also, work in a well-ventilated area away from open flames (such as the pilot lights on stoves, furnaces, and water heaters).

While you're shopping for a filter mask, pick up a good pair of safety glasses, and wear them whenever you use a motor tool.

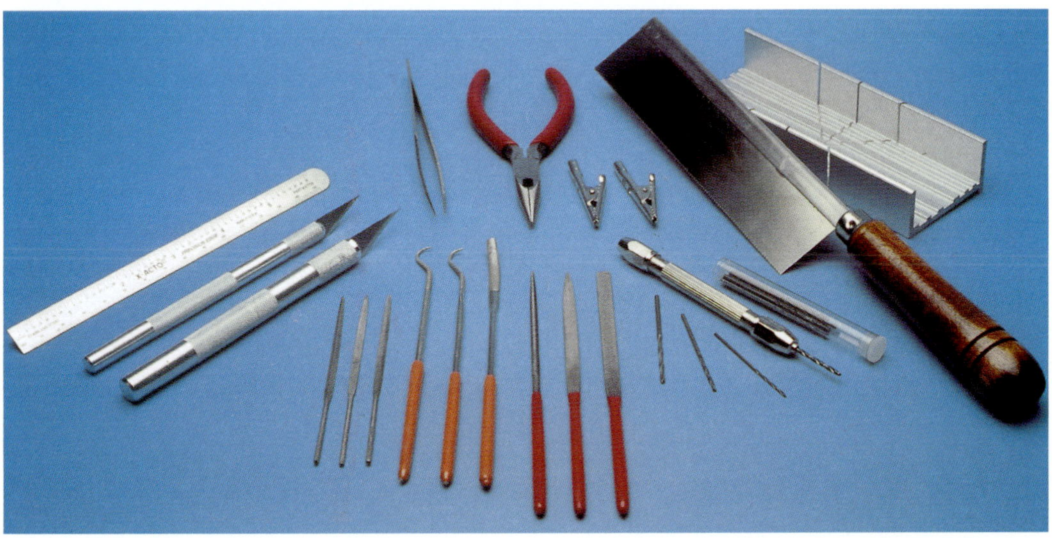

Having a good variety of hand tools on your workbench makes any project easier. A pin vise, razor saw, pliers, tweezers, and a variety of files should be on every workbench.

Lots of different glues are available at the hobby shop and hardware store, but which one should you use on your model? Actually, all these glues are good for modeling, albeit for different reasons. Here's a rundown of which glues are best for which uses:

1: White glue Don't laugh—white glue is great for installing clear parts, such as windows and canopies. It won't mar them like solvent-based glues.

2: Epoxy Epoxy is a two-part glue that's very strong and well-suited for times when strength is needed, or when dissimilar materials must be joined. Different epoxies cure at different rates (from five minutes to more than an hour), so pick the one that's most appropriate for your project.

3: Super glues Super glue is a strong adhesive that's great for general assembly. It's available in a variety of viscosities and cure times. Applying super glue "kicker" to the joint after the parts are put together will cause the glue to cure instantly and lock the parts in place. Super glue can bond skin instantly, so always have a bottle of debonder handy—look for it near the glue in the hobby shop. Also, super glue can "frost" and ruin the surface of clear plastic parts, so attach them to your models with white glue or epoxy.

4: Tube glue Tried-and-true solvent-based tube glue is still a good choice for general assembly. It only works on plastic, though, so if you're planning to use other materials (such as cast-resin or metal parts) you'll need to use another glue.

5: Liquid cement Like tube glue, solvent-based liquid cements provide strong bonds between parts. A fine-point paintbrush will help you apply the glue accurately.

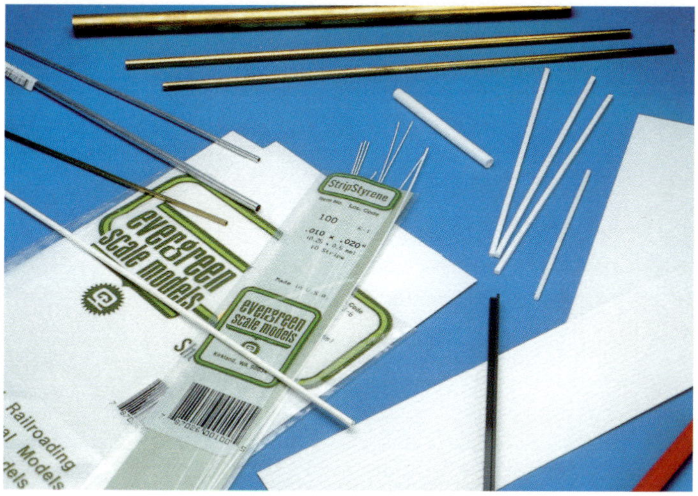

If you've looked at any of the other chapters in this book, you know that I use a lot of plastic, aluminum, and brass parts to detail my models. Keep a variety of styrene sheet, strip, and rod on hand, as well as a few aluminum and brass shapes. They'll make adding detail and solving little fit problems a whole lot easier.

For the cost of a couple of plastic kits, you can buy an inexpensive rechargeable motor tool like this one. Regardless of what kind of motor tool you buy, make sure it has a low-speed setting that's no faster than 5,000 rpm or so. If you use a tool that operates faster than that on a plastic model, the tool will melt the plastic instead of cutting or grinding it. Pick up a few drill and grinding bits, too, and you'll be ready to tackle more advanced projects.

Remember, the studio models were often detailed with parts from off-the-shelf plastic kits. The detail part you're looking for may be hiding in one of those unbuilt car or tank kits on your shelf!

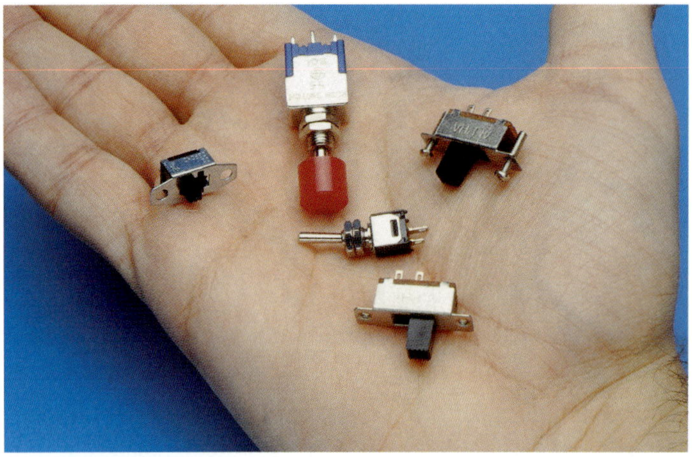

There's a huge variety of switches available if you're thinking about adding lights or other electrical details to your models. All of these examples came from Radio Shack, and the smallest are easy to hide on (or in) most models.

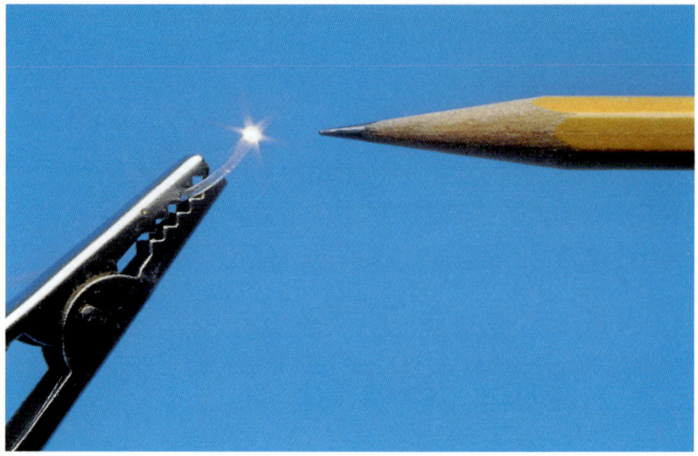

Fiber optic is extremely flexible; it can be used to pipe light into (and through) confined spaces. I use it to add engine lights to an Imperial TIE fighter elsewhere in this book.

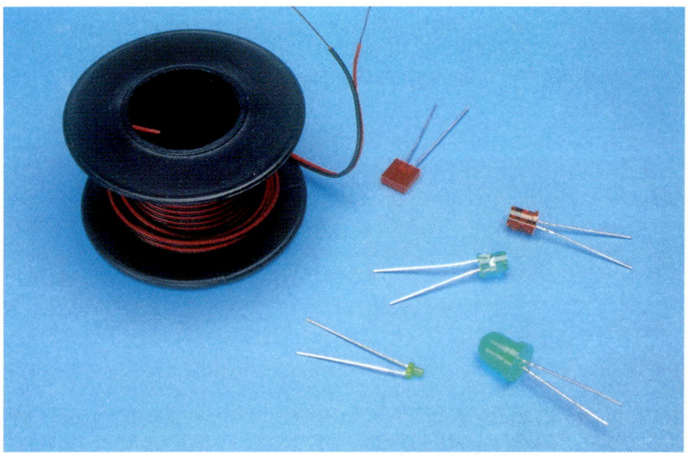

Light-emitting diodes (LEDs) are an easy, inexpensive way to light a model. They're available in a variety of shapes, sizes, and colors, and can be powered by a couple of flashlight batteries.

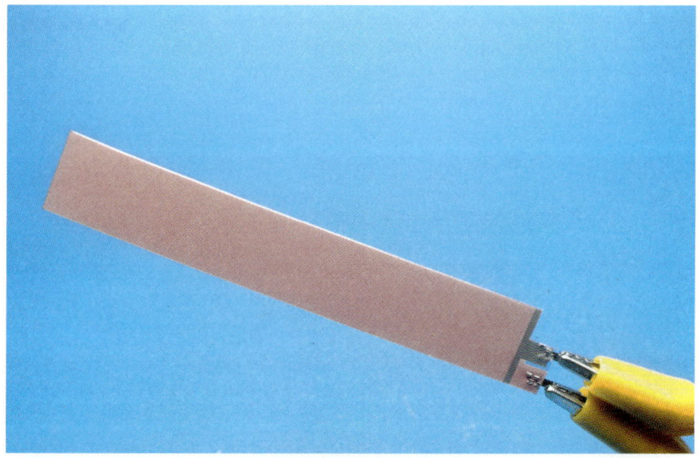

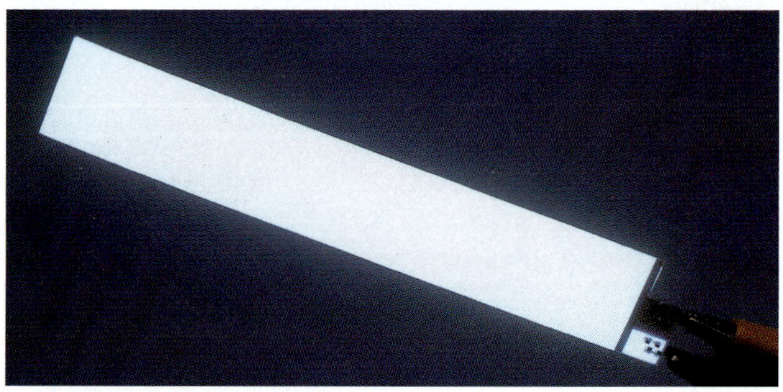

LightSheet is a thin, flexible plastic film that glows like neon when powered by a 9-volt battery. Available in blue and white, LightSheet generates no heat and can be cut into almost any shape. I used it to model the cockpit and engine lights in my Millennium Falcon kit.

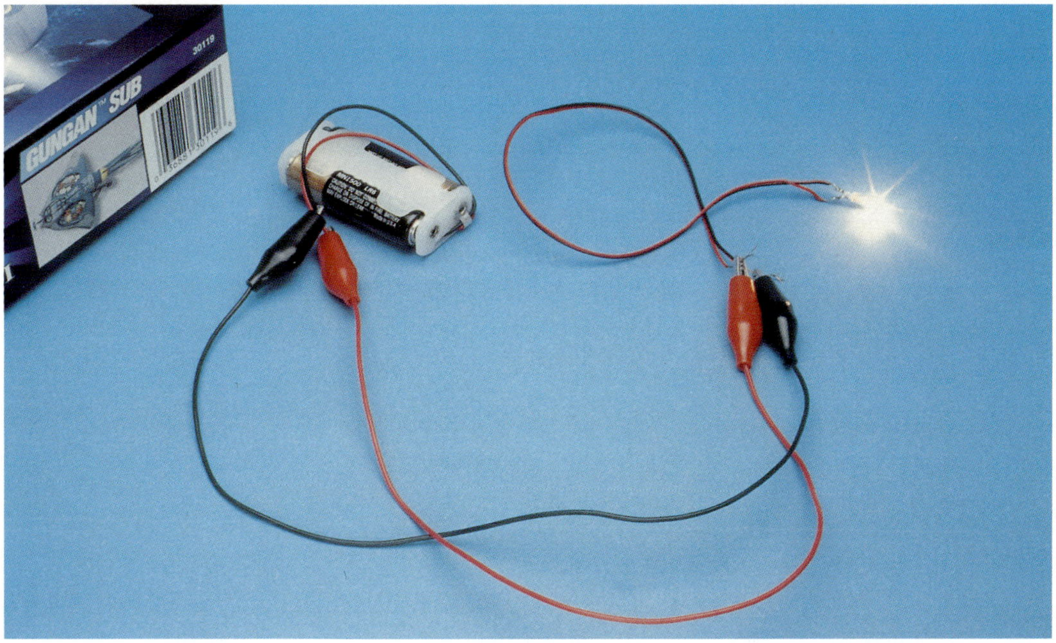

If you're going to do any electrical work at all, pick up a set of test leads. These color-coded leads, usually sold in sets of ten or so, have alligator clamps at each end. They'll allow you to experiment and try out different wiring configurations quickly before you assemble things permanently. Look for a set at electronics-supply stores.

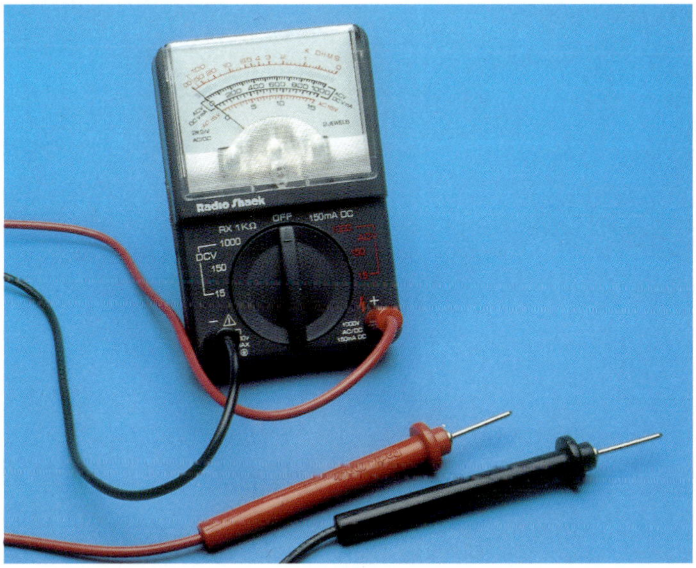

While you're at the electronics-supply store, pick up an inexpensive multimeter like this one. It'll provide a variety of uses, including the ability to determine the exact voltage of your power source, which can keep you from accidentally damaging delicate circuitry and components.

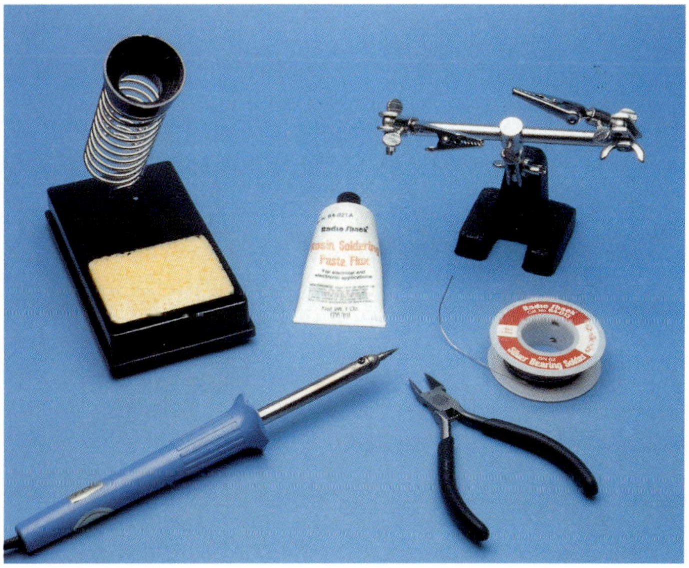

Soldering is really the only way to ensure good electrical connections if you add lights or motors to your models. It's not as hard as you might think, and most of the tools, such as the pencil-type soldering iron and "extra-hands" parts holder, are fairly inexpensive.

Thankfully, there are plenty of references for building *Star Wars* models. Books, magazines, trading cards, and boxtop photos all come in handy, and "doing research" is always a good excuse to watch the movies again!

Mike Fulmer profile

You may not recognize his name, but you'll almost certainly recognize his models. A master model builder, Mike Fulmer is responsible for some of the most famous special-effects models in movie history. After more than ten years at Lucasfilm's Industrial Light & Magic (ILM) production house, his work has been featured in some of the highest-grossing films of all time, including *The Empire Strikes Back, Raiders of the Lost Ark, Return of the Jedi, Back to the Future, E.T.,* and *Star Trek II: The Wrath of Khan.*

Ironically, Mike didn't set out to be a professional model builder. After serving with the Marines in Vietnam, Mike returned to the states and worked at a variety of jobs, including as an ironworker and pipe fitter. The son of an Air Force pilot, Mike had always been interested in model building, particularly scratchbuilt aircraft, and sometimes sold his models to collectors.

While working as a welder in Oregon in the late 1970s, Mike built an air racer model for a collector in San Francisco. When he delivered the model, the collector, who had retired from television work in Hollywood, encouraged Mike to show it to the staff of a film production company that was just moving into new facilities in San Rafael. Reluctantly, Mike agreed and, armed with the finished model, had an informal interview. Afterward he dropped off the model and drove back to Oregon.

The company, George Lucas' Industrial Light & Magic, had just moved north from Van Nuys, fresh from the success of *Star Wars.* A few weeks later Mike received a letter asking him to join the staff, which was gearing up to start production on the second *Star Wars* film, *The Empire Strikes Back.* After a little debate, Mike decided to leave his job as a steelworker and take the job. He was a little uncertain about what exactly he would be doing, though—Mike had never seen *Star Wars.*

It was a hectic time at ILM. The company was building a new special-effects facility from scratch and working on *Empire* at the same time, all under Lucas' tight preproduction security. At one point Mike took a model into a restroom just to get away from the construction noise long enough to do some hand-painting.

One of his first assignments ended up being one of the icons of the *Star Wars* universe—Han Solo's Millennium Falcon. Smaller than the original 4-foot model used for the first film, the model enabled the crew to film more complex sequences, including a breakneck race through the asteroid field. Mike ended up building the Falcon in several scales to accommodate different shots.

Despite his science-fiction beginnings, Mike's favorite project is Steven Spielberg's *Always,* in which he was able to get back to his aircraft roots. The film's scratchbuilt A-26 and Catalina firebomber models had wingspans up to 20 feet. All of the ships' systems, from the control surfaces to the engine speed, were controlled remotely during the film's firefighting sequences.

> Modeling as a hobby is usually a part-time affair, and the subject model is of interest to the builder, kind of a labor of love .... High-end model work is a serious business.

After working on more than 30 films and commercials, Mike left full-time work at ILM. Although he still does special-effects work, he now has more time to concentrate on other interests.

An in-progress 1/15 scale World War II fighter rests on the workbench in his home workshop north of San Francisco, bound for a private collector. All of the model's panels were hand-formed over a wooden buck and assembled over a photoetched airframe. The plane's control surfaces move when I reach in the cockpit and push on the 4″-long control stick.

Also on the bench are several prototype models for electric train giant Lionel. Brought into the company by Neil Young in the mid-1990s, Mike designed several models for the company's Route 66-themed line of trains and accessories.

Not all of Mike's projects will fit on the workbench, though. A freshly completed full-size 1932 Ford street rod rests under a car cover across the workshop. Mike has two other larger workshops for projects where more space or security is required.

"Over the years, quite a few modelers, usually talented hobbyists, have asked me how they could enter the film business. That question's always a handful to answer," Mike says. "There's a wide gap between modeling as a hobby and modeling as a professional. Modeling as a hobby is usually a part-time affair, and the subject model is of interest to the builder, kind of a labor of love.

"You have to leave all that behind when you cross into the professional arena," he continues. "I've known a few talented modelers who took the high-level path, and couldn't handle the 12-hour days, seven-day weeks, and the push to produce projects that held no special interest other than a paycheck. High-end model work is a serious business."

MPC and AMT/Ertl Star Wars kits

The first pages of MPC's 1978 catalog announced the company's new line of kits. "MPC proudly announces the first introductions of a superb new kit series based on the epic film: *Star Wars*," the copy read. "Everything possible has been done to execute this series in a manner consistent with the quality of this remarkable film!"

For all the excitement, only four kits were included in the pages that followed, two figures (R2-D2 and C-3PO) and two spacecraft (Darth Vader's TIE Fighter and Luke Skywalker's X-Wing Fighter). Needless to say, other kits followed the original four in huge numbers (even after MPC was acquired by AMT/Ertl in 1985), and the series would become one of the largest catalogs of licensed kits in the history of plastic models. What follows is a list of all the *Star Wars* model kits produced (or reissued) by both companies, including their catalog numbers.

KIT NAME	CATALOG NUMBER
Three-Piece Gift Set	No. 8912 (AMT)
AT-AT	No. 1918 (MPC), no. 1929 (MPC), no. 8918 (AMT)
AT-AT Structor	No. 1902 (MPC), no. 6036 (AMT)
SnapFast AT-ST	No. 1976 (MPC)
AT-ST Structor	No. 1903 (MPC), no. 6029 (AMT)
AT-ST Walker	No. 1976 (MPC), no. 8734 (AMT)
AT-ST Mirr-A-Kit	
SnapFast A-Wing Fighter	No. 1973 (MPC), no. 8933 (AMT)
Anakin's Podracer	No. 30122 (AMT)
SnapFast B-Wing Fighter	No. 1974 (MPC)
Battle on Ice Planet Hoth Action Scene	No. 1922 (MPC), no. 8743 (AMT)
C-3PO	No. 1913 (MPC), no. 1935 (AMT)
C-3PO Structor	No. 1901 (MPC)
Cantina Action Scene	No. 8205 (MPC)
Darth Vader Action Kit	No. 1921 (MPC)
Darth Vader Van	No. 3209 (MPC)
Darth Vader plastic figure	No. 1916 (MPC), no. 8154 (AMT)
Darth Vader's TIE Fighter	No. 1915 (MPC), no. 8916 (AMT)
Darth Vader's TIE Fighter Plus Pack	No. 8432 (AMT)
Darth Vader vinyl figure	No. 8784 (AMT)
Darth Vader's TIE Fighter Flight Display	No. 8275 (AMT)
Snap Fast Death Star	No. 8193 (AMT)
Encounter with Yoda Action Scene	No. 1923 (MPC), no. 8263 (AMT)
Shadows of the Empire Emperor Palpatine vinyl figure	No. 8258 (AMT)
Shadows of the Empire Xizor's Virago	No. 8377 (AMT)
Shadows of the Empire Xizor vinyl figure	No. 8256 (AMT)
Han Solo vinyl figure	No. 8785 (AMT)
SnapFast Gungan Sub	No. 30119 (AMT)
Imperial TIE Fighters	No. 8438 (AMT)
Imperial Star Destroyer	No. 1926 (MPC), no. 8915 (AMT)
Fiber-Optic Star Destroyer	No. 8782 (AMT)

KIT NAME	CATALOG NUMBER
Jabba the Hutt Throne Room Action Scene	No. 1928 (MPC), no. 8262 (AMT)
Luke Skywalker Van	No. 3210 (MPC)
Luke Skywalker vinyl figure	No. 8783 (AMT)
Millennium Falcon	No. 1925 (MPC), no. 1933 (MPC), No. 8917 (AMT)
Millennium Falcon Cut-Away	No. 8789 (AMT)
Die-cast Naboo Fighter	No. 30130 (AMT)
SnapFast Naboo Fighter	No. 30117 (AMT)
R2-D2 Van	No. 3211 (MPC)
R2-D2	No. 1912 (MPC), no. 1934 (AMT)
Rancor with Gamorrean Guard	No. 8171 (AMT)
Rebel Base Action Scene	No. 1924 (MPC), no. 8735 (AMT)
Shuttle Tydirium	No. 1920 (MPC), no. 8733 (AMT)
Shuttle Tydirium Mirr-A-Kit	
Slave I	No. 1919 (MPC), no. 8768 (AMT)
Snowspeeder	No. 1917 (MPC), no. 8914 (AMT)
Speeder Bike	No. 1927 (MPC), no. 8928 (AMT)
Speeder Bike Flight Display	No. 6352 (AMT)
Speeder Bike Mirr-A-Kit	
SnapFast TIE Interceptor	No. 1972 (MPC), no. 8932 (AMT)
TIE Interceptor Mirr-A-Kit	
Trade Federation Battle Tank	No. 30123 (AMT)
SnapFast Trade Federation Droid Fighters	No. 30118 (AMT)
Trade Federation STAP with Battle Droid	No. 30124 (AMT)
Animated X-Wing Fighter	No. 8585 (AMT)
X-Wing Fighter	No. 1914 (MPC), no. 1930 (MPC), No. 8918 (AMT)
X-Wing Fighter Flight Display	No. 8788 (AMT)
X-Wing Fighter Plus Pack	No. 8433 (AMT)
X-Wing Fighter Mirr-A-Kit	
SnapFast X-Wing Fighter	No. 1971 (MPC), no. 8931 (AMT)
SnapFast Y-Wing Fighter	No. 1975 (MPC), no. 8934 (AMT)
Y-Wing Fighter Mirr-A-Kit	

Sources

Bare-Metal Foil
P.O. Box 82
Farmington, MI 48332
Tel.: 248-477-0813
Fax: 248-476-3343
E-mail: baremetal@mediaone.net
Web: www.bare-metal.com

Evergreen Scale Models
18620-F 141st Ave. N.E.
Woodinville, WA 98072
Tel.: 425-402-4918 or 877-376-9099
Fax: 425-402-7948

Falcon Kits
E-mail: sales@falconkits.com
Web: www.falconkits.com

LightSheet Systems
319 Main Dunstable Rd.
Nashua, NH 03062-1906
Tel.: 603-595-7146
Fax: 603-595-9609
E-mail: TrekFX@aol.com

Micro-Mark Tools
340 Snyder Ave.
Berkeley Heights, NJ 07922
Tel.: 908-464-6764
Fax: 908-665-9383
E-mail: info@micromark.com
Web: www.micromark.com

Miniatronics
561-K Acorn St.
Deer Park, NY 11729
Tel.: 800-942-9439
Fax: 516-242-7796
E-mail: miniatronics@miniatronics.com
Web: www.miniatronics.com

Pacer Technology
9420 Santa Anita Ave.
Rancho Cucamonga, CA 91730
Tel.: 800-538-3091 or 909-987-0550
Fax: 909-987-0490
Web: www.pacertech.com
E-mail: sales@pacertechnology.com

Plastruct
1020 South Wallace Place
City of Industry, CA 91748
Tel.: 800-666-7015
Fax: 626-965-2036
E-mail: plastruct@plastruct.com
Web: www.plastruct.com

Testor
620 Buckbee St.
Rockford, IL 61104
Tel.: 800-837-8677 (U.S. and Canada only)
Fax: 815-962-7401
Web: www.testors.com